Climate of Denial

CLIMATE OF DENIAL

*Darwin,
Climate Change,
and the Literature of
the Long Nineteenth Century*

Allen MacDuffie

Stanford University Press
Stanford, California

Stanford University Press
Stanford, California

©2024 by Allen MacDuffie. All rights reserved.

No part of this book may be reproduced or transmitted in any form or by any means, electronic or mechanical, including photocopying and recording, or in any information storage or retrieval system, without the prior written permission of Stanford University Press.

Printed in the United States of America on acid-free, archival-quality paper.

Library of Congress Cataloging-in-Publication Data
Names: MacDuffie, Allen, author.
Title: Climate of denial : Darwin, climate change, and the literature of the long nineteenth century / Allen MacDuffie.
Description: Stanford, California : Stanford University Press, 2024. | Includes bibliographical references and index.
Identifiers: LCCN 2023046212 (print) | LCCN 2023046213 (ebook) | ISBN 9781503638938 (cloth) | ISBN 9781503639546 (paperback) | ISBN 9781503639553 (ebook)
Subjects: LCSH: Darwin, Charles, 1809-1882—Influence. | English literature—19th century—Themes, motives. | Denial (Psychology) in literature. | Climatic changes—Psychological aspects.
Classification: LCC PR468.D44 M33 2024 (print) | LCC PR468.D44 (ebook) | DDC 820.9/3552—dc23/eng/20240125
LC record available at https://lccn.loc.gov/2023046212
LC ebook record available at https://lccn.loc.gov/2023046213

Cover design and color overlay: Daniel Benneworth-Gray
Cover art: Henrique Alvim Corrêa, *Martian Viewing Drunken Crowds*, 1906, H. G. Wells's novel *The War of the Worlds*, published by L. Vandamme & Cie in a French translation by HD Davray.

For Cecile MacDuffie and Nadia Vine, with love

CONTENTS

Acknowledgments   ix

INTRODUCTION
We Have Never Been Darwinian   1

1   Forms of Denial   18

2   The Denial of Darwin   51

3   Denial in the First Person   97

4   George Eliot and Free Indirect Denial   131

5   Virginia Woolf and the Ends of Denial   181

CONCLUSION
Full Disclosure   217

Notes   237

Bibliography   259

Index   273

ACKNOWLEDGMENTS

ONE OF THE THEMES OF this book is how hard it is to acknowledge certain things. In this case, the only difficulty I face is finding a way to express adequately how deeply grateful I am to the many people who helped make it possible.

I begin with the members of the Vcologies Working Group, where, in 2017, I workshopped the short paper that would one day become this book. Special thanks to Elizabeth Carolyn Miller, whose invitation to develop that paper into an article for a special issue of *Victorian Studies* led me to begin researching these questions more deeply. Her enthusiasm for the project and keen editorial eye were invaluable guides during the crucial early stages.

Earlier versions of sections of Chapters 2 and 3 appeared in the article "Charles Darwin and the Victorian Pre-History of Climate Denial," which was published by *Victorian Studies* in 2018. Many thanks to the editorial team there.

Major parts of this book were written during a research leave in 2019–20 when I was an Early Career Fellow at the University of Pittsburgh Humanities Center. I'm grateful to Jonathan Arac, Dan Kubis, David Marshall, Gayle Rogers, Lisa Brush, and all of the faculty fellows who responded with rigor and generosity to early drafts.

Thanks to Pauline Strong and my fellow faculty members at the University of Texas at Austin Humanities Institute, where I spent an immensely rewarding semester in conversation with colleagues across the university. It was a sustaining experience to be a part of such a vibrant, collegial community at a moment in the pandemic when community was especially hard to come by.

I'm grateful to Sam McKegney and the members of the Queen's University English Department, where I spent an invigorating year finishing a full draft of this manuscript. The faculty and graduate student participants at the Research Forum were an ideal audience whose generous attention and incisive questions came at the perfect moment in the process.

Thanks to all my colleagues at the University of Texas who have made this the best possible place to call my academic home: Daniel Birkholz, Samuel Baker, J. K. Barret, Edward Carey, Elizabeth McCracken, Gretchen Murphy, Coleman Hutchison, Heather Houser, Carol MacKay, Lisa Moore, Hannah Wojciehowski, Julia Mickenberg, Elizabeth Cullingford, John González, Patricia García, Mia Carter, Douglas Bruster, Martin Kevorkian, Janine Barchas, Elizabeth Scala, Lance Bertelsen, Neville Hoad, Oscar Cásares, Domino Perez, Brian Doherty, Alan Friedman, Wayne Rebhorn, Phillip Barrish, Samantha Pinto, Ann Cvetkovich, David Kornhaber, James Cox, George Christian, Elizabeth Richmond-Garza, John Pipkin, Chad Bennet, Donna Kornhaber, and Julie Minich. Rest in peace Don Graham, Brian Bremen, Wayne Lesser, Tom Whitbread, and James Loehlin.

Many thanks to my letter writers for all the support over the years: Cannon Schmitt, Chip Tucker, Joe Childers, and Liz Miller. I don't know why I'm always afraid to ask when you're always so nice about it.

I'm deeply grateful to the two anonymous reviewers of this book, both for the generosity with which they read the manuscript and the skeptical pressure they applied to it. Their reports were absolutely indispensable—they were the decisive guides I needed through the last crucial stages of writing.

So many thanks to Caroline McKusick at Stanford University Press for her initial belief in the project and her unfailingly acute editorial eye throughout the entire process.

Shout out to sometime housemates and constant friends on the text thread: Aman Garcha, David Kurnick, and Gerard Cohen-Vrignaud.

Thanks to old friends, far and near: Curtis Brown, Judd Spray, John Grennan, Peter D'Amico, Peter Saharko, Mike McCarthy, Kevin Birmingham, Daniel Yates, and Nathan Townsend.

Thanks to Nick, gentlest of creatures and truest of friends. And to Chip Mitchell, patiently drumming his fingers somewhere.

Thanks to my family: Brent Vine, Olga Yokoyama, George Taylor, Susan Mormile, Elizabeth Reger, Kate Lozano, Santiago Lozano-Calderon, and Santi, Sebastian, Mateo, Nicky, and Ana Lucia Lozano.

To Richard Keegan: I hope the bees gave you my message.

To Mom and Dad, who will pretend they're going read this book, even though we all know they won't.

And finally to Cecile MacDuffie, Nadia Vine, and Vera Vine—always here with me, even when I can't be there with them.

Climate of Denial

INTRODUCTION

WE HAVE NEVER BEEN DARWINIAN

"EVOLUTION," WRITES ANNIE DILLARD IN *Pilgrim at Tinker Creek* (1974), "loves death more than it loves you or me. This is easy to write, easy to read, and hard to believe. The words are simple, the concept clear—but you don't believe it, do you?"[1] Dillard recalls a brief but memorable moment from the third chapter of *The Origin of Species*, where, also writing in personal terms, Charles Darwin describes the same dilemma: "Nothing is easier than to admit in words the truth of the universal struggle for life, or more difficult—at least I have found it so—than constantly to bear this conclusion in mind."[2] If the difficulty these writers describe can be understood as a form of denial, it is not what we usually have in mind when we talk about "science denialism." At issue here is not the open rejection or deliberate obfuscation of evolutionary science but the acknowledgment of an emotional obstacle, some kind of internal resistance to fully accepting the picture of nature it paints. Darwin's word is "admit," which he tells us is easy enough to do verbally. But "admit" of course has another sense—to allow inside—and this is where things get "more difficult." "Admission," we might say, is denial's mirror image, its more obliging twin. Both words are cleaved semantically between a mere speech act of attestation (or gainsaying) and a deeper, more psychologically fraught acceptance (or rejection) of whatever uncomfortable, alienating, dislocating, or aversive thing is under discussion.

Darwin's metaphor of "bearing" in mind—so common, admittedly, as to almost no longer count *as* a metaphor—suggests something burdensome about these ideas when allowed inside and carried around "constantly."³ It was a burden Darwin sometimes wished he could relinquish.

For both Dillard and Darwin, the problem is not about *whether* to accept certain natural realities as true but how to live with them. What both writers express is a version of that special category of denial known as disavowal. Disavowal—often glossed as "knowing and not knowing at the same time"—is a complex concept, the various terminological issues surrounding which will be taken up later. For now, I would point to the ways in which it has increasingly become a part of the popular and scholarly conversation about the response to anthropogenic climate change. The sociologist Kari Marie Norgaard describes a "failure to integrate" certain forms of scientific knowledge into everyday experience, which produces a "double reality." "In one reality," she writes, there is "the collectively constructed sense of normal everyday life. In the other reality . . . the troubling knowledge of increasing automobile use, polar ice caps melting, and the predictions of future [extreme] weather scenarios."⁴ The sociologist Stanley Cohen calls this "implicatory denial," to suggest a disconnect between accepted knowledge and the implications that knowledge holds for how one should live.⁵ The French psychoanalyst Octave Mannoni's famous line "I know well, but all the same . . ." (*Je sais bien, mais quand même . . .*) captures the sense of disconnect: the split in the disavowing subject that is accompanied (crucially) by the subject's own self-awareness *of* that split.⁶ In this book, I lean on Naomi Klein's term "soft denial" because of the way it pairs with its extremist counterpart "hard denial" to suggest not only something of the range of evasive responses but also their interarticulation.⁷ The expansion of the idea of denial to include many who actually *accept* climate science speaks to the much-observed gap between widespread knowledge of the crisis and the lack of urgency on the part of policy makers and the public to respond in a way commensurate with the nature and scope of the problem. As Bruno Latour puts it in *Facing Gaia*, "We can't say that we didn't know. It's just that there are many ways of knowing and not knowing at the same time."⁸

The tactics of hard deniers are all too familiar to those of us watching with horror as clouds of misinformation about evolution and climate science continue to spew from right-wing media outlets, corporate-funded think tanks, and a rogues' gallery of various cynics, grifters, and con artists. But if the hard denial of evolution persists well beyond the nineteenth century into the present, Dillard's musings in *Pilgrim at Tinker Creek* illustrate that soft denial clearly does as well, even among those who are actually *trying* to come to terms with it. "I ought to keep a giant water bug in an aquarium on my dresser, so I can think about it," she writes. "We have brass candlesticks in our houses now; we ought to display praying mantises in our churches. Why do we turn from the insects in loathing?"[9] Dillard's book is in many ways a record of an entire year in which she works to "bear constantly in mind" the basic natural processes upon which Darwin's theory depends, discovering along the way that she could still use some imaginative aids around the house. Because, of course, the answer to her question about the pull of denial— why we turn away "in loathing"—has everything to do *with* those houses and churches and brass candlesticks and any number of other cultural practices, commodities, and features of the built environment that function— implicitly or explicitly—to enforce a spurious sense of difference between human life and the world of "nature." The reason churches don't have statues of praying mantises and people don't keep water beetles on their dressers is that such objects would undermine the whole point of such spaces and artifacts, which is to make possible a reality where one can feel, for a time, that insects don't really matter. Don't, in some sense, *exist*. Dillard's journey in *Pilgrim at Tinker Creek* is framed as an individual soul's solitary struggle to confront the natural world, but in some ways what it really documents is the struggle of an individual soul trying to free herself from a culture of denial.

What Dillard makes clear, in her many moments of "turning away in loathing" from unsettling natural scenes, is the divide between accepting something to be true and real and fully assimilating it into one's sense of identity or experience of the world. In her last book, *The Weather in Proust*, Eve Kosofsky Sedgwick usefully frames this in terms of the split between "reality and realization," defining the former as propositional forms of knowledge and the latter as "process and practice"—a more comprehensive "nondual" striving toward the integration of the emotions, the senses, and the understanding at once.[10] Sedgwick is talking about her struggle to come to terms with her own terminal illness and the difference between acknowledging her mortality as a fact, on the one hand, and actually "understanding it as real," on the other.[11] Her short essay doesn't mention the climate crisis or anything like it, but in its critique of an ingrained subject-object binarism that is partly responsible for the "shrunken impoverishment of any Western psychology of knowledge and realization," it speaks to recent deconstructive work on ecology by Elizabeth Grosz and Timothy Morton (more about which later) and gets at the reasons why climate or evolutionary science can be treated as entirely real while nevertheless remaining stubbornly "unrealized" for most people, most of the time.[12] "How normal it is," she writes, "for realization to lag behind knowledge for months or eons"—or, to take the middle scale perhaps most appropriate to the climate crisis, for decades.[13] Of course, "normal" here means something more like "normalized"—taken as normal even if it is maladaptive or destructive. Or entirely mad.

Thus, it is a short step from *The Weather* to the climate. The neurobiologist Janis Dickinson has recently argued that the denial of climate change is really an expression of the denial of mortality with which Sedgwick struggles.[14] For the anthropologist Ernest Becker, upon whose work Dickinson draws, the denial of death produces what he calls "immortality projects": culturally generated fantasies that keep anxious thoughts of extinction at bay at the steep cost of fundamentally misrepresenting the nature and reality of our animal existence to ourselves. These "projects" can be religious of course (where personal immortality is explicitly on offer), but they can also take the form of discourses of entrepreneurial success, myths of heroic individualism, or utopian ideologies of both left- and right-wing varieties. For Becker, the "New Universal Immortality Ideology" is, at its root, all about wealth accumulation and the fantasy that money can bestow power and permanence, can rescue its possessor from fears of insignificance and oblivion. Though Becker doesn't specifically discuss disavowal, clearly "immortality ideology" at least partly falls under this category—excepting those who plan to freeze their heads in cryo-storage, even rich people know you can't buy your way out of death. Becker's argument is that the worship of money represents not so much an outright denial of creaturely existence but a transmutation of it into a sense of immediate, palpable control over one's environment: "Man succumbs easily to the temptation of created life, which is to exercise power mainly in the dimension in which he moves and acts as an organism. The pull of the body is so strong, lived experience is so direct; the 'supernatural' is so remote and problematic, so abstract and intangible."[15] Becker's line of argument has been taken up more recently in the work of the geographer Yi-Fu Tuan, who argues that all of human culture is, at heart, fundamentally about denying the body: "Culture is the totality of means by which I escape from my animal state of being."[16] Struggling to maintain a false belief in one's exceptional status is the thing that makes one exceptional, even though, of course, it also doesn't. As the novelist John Fowles puts it, "We are all in a flight from the real reality. That is a basic definition of *Homo sapiens*."[17] "Escapism," argues Tuan, "is human—and inescapable."[18] Thus, on some level, to be human is to refuse to accept being human. Reality is produced in the efforts to flee it.

And yet, of course, such universalizing accounts themselves tend to obscure—we might even say "deny"—the social, historical, and political dimensions of denialism: its roots in capitalism and the commodity form; in the numerous contradictions of the Enlightenment project of Western humanism; in empire and all its attendant mystifications and erasures. Although "denial" often gets used to identify an individual psychological shortcoming or moral failing, it is also, and more importantly for our purposes in this book, a complex, organized, well-funded cultural production that trades upon, as it generates, elaborate fantasies of various kinds (security, innocence, uniqueness, agency, superiority, virtue). This is ignorance not as lack, or shortcoming, or failure, but as a product intrinsic to the logic of capitalism, because it is needed to disguise its rapaciously anti-human operations and its fundamental incompatibility with equality, democracy, human rights, and all of the other putative ideals of the liberal humanist project. Indeed, as we are discovering, capitalism's fundamental incompatibility is with life itself. So while it may be entirely true that *Homo sapiens* is hardwired for denial, that we are somehow not rigged, neurally speaking, to dwell for long in remote, transpersonal vistas of space and time; while it may be the case that we have, by and large, not evolved to consistently extend our radius of sympathetic attention much beyond our own short-term needs and those of our own family or kin group or tribe, it is also true that advanced consumer capitalism has become expert at exploiting and supercharging those predispositions to denialism; at fortifying and celebrating (fortifying by celebrating) the felt experience of individual centrality and significance; at fueling fantasy life and then inviting us at every turn to push whatever is unwanted—which it has also worked tirelessly to define—out of mind.

The unresolved contradiction in the cultural uptake of evolutionary ideas—a widespread acceptance of Darwinian science accompanied by a refusal of its implications—has been noted in recent works by Bruno Latour, Adam Phillips, Donna Haraway, and many other critics we'll encounter in the following pages. Latour notes that many card-carrying "Darwinians," including Richard Dawkins, are guilty of this: "It is safe to say that one hundred and fifty years after his discoveries, the full originality of Darwin's thought has still not been absorbed by public consciousness. I am not alluding here to the masses of results and models obtained by evolutionary biologists, but to the metaphysical consequences of evolutionary theory. The problem is that the full originality of Darwin's understanding of the world is obfuscated not only by so-called creationists but in part also by neo-Darwinians."[19] Latour locates the obfuscation of Darwinian ecology at the root of our inability to adequately respond to, or even just successfully imagine, the magnitude of the environmental crisis we face. A similar critique was frequently leveled by the paleontologist Stephen Jay Gould, who argued that the idea of progress produced, and was produced by, a form of denialism that implicitly or explicitly re-encoded hierarchy and direction in the natural order despite the profoundly anti-teleological thrust of Darwin's entire project.[20] "We are still not ready for the Darwinian revolution," he writes, because we refuse to "own the plain implications of evolution for life's nonpredictable nondirectionality."[21] "Owning" those implications would mean accepting that humanity is neither point nor pinnacle of organic life; that we differ from non-human animals only in degree, not in kind; that we do not and cannot "transcend" or "subdue" nature but remain fully embedded within and dependent upon it; that the earth, not having being made for us, might change, or be changed, in ways that make it no longer hospitable to us. It would mean, as he puts it in *Ever Since Darwin*, once and for all letting go of the "fallacious equation of organic evolution with progress," which "remains a primary component of our global arrogance, our belief in dominion over, rather than fellowship with, more than a million other species that inhabit our planet."[22] Pretty obvious, but Gould contends that such implications are not taken "seriously" because, if they were, there would be a radical rethinking of so many of the most basic categories of human meaning and modes of social and political organization.[23] Gould's target is not so much those who openly deny the facts of evolution but self-described Darwinians, including his own colleagues in evolutionary biology, who practice a more cagey kind of disavowal, which he calls "spin" and "special pleading."[24]

One biologist who gets singled out for special criticism is E. O. Wilson, who surely would have objected to the description of his work on Darwin as "spin." And yet, interestingly, Wilson also somewhat cryptically admits to practicing the kind of denialism Gould lays at his door. In the final passage of *On Human Nature*, he mounts a defense of the progressive vision of modern science:

> In the spirit of the enrichment of the evolutionary epic, modern writers often summon the classical mythic heroes to illustrate their view of the predicament of humankind: the existential Sisyphus, turning fate into the only means of expression open to him; hesitant Arjuna at war with his conscience on the Field of Righteousness; disastrous Pandora bestowing the ills of mortal existence on human beings; and uncomplaining Atlas, steward of the finite Earth. Prometheus has gone somewhat out of fashion in recent years as a concession to resource limitation and managerial prudence. But we should not lose faith in him. Come back with me for a moment to the original, Aeschylean Prometheus:
>
> *Chorus*: Did you perhaps go further than you have told us?
> *Prometheus*: I caused mortals to cease foreseeing doom.
> *Chorus*: What cure did you provide them with against that sickness?
> *Prometheus*: I placed in them blind hopes.
>
> The true Promethean spirit of science means to liberate man by giving him knowledge and some measure of dominion over the physical environment. But at another level, and in a new age, it also constructs the mythology of scientific materialism, guided by the corrective devices of the scientific method, addressed with precise and deliberately affective appeal to the deepest needs of human nature, and kept strong by the blind hopes that the journey on which we are now embarked will be farther and better than the one just completed.[25]

The philosopher Charles Taylor argues that Wilson displays "a sublime indifference to inconsistency" here and throughout this book, insofar as he argues, on the one hand, for a reductive view of human moral life as fundamentally irrational, emotive, and unconscious (because hardwired by thousands of generations of natural selection) and, on the other, for a vision of heroic human self-fashioning aimed at the progressive loosening of the bonds of nature.[26] But it seems to me that what we have here is not so much indifference to inconsistency as a strategic embrace of it. To subscribe to "scientific humanism" (as Wilson calls it) is to posit a world in which the human is decentered in reality but is still kept centered in the story of "reality" that we all agree to tell ourselves. It is what he calls the "evolutionary epic . . . probably the best myth we will ever have."[27] Wilson is not so much declaring his belief in the "myth" of human progress as arguing for the importance of suspending *disbelief* in it.[28] But note that, in order to do this, other concerns must be minimized or ignored—the focus on "resource limitation and managerial prudence," for example. That is, what gets backgrounded are precisely the kinds of ecological pressures and limits that might cast doubt on Promethean ambitions. The validity of such concerns is not disputed; instead, they are simply pushed to one side, consigned to a different mythological framework we are deciding not to adopt.

Taylor calls Wilson's "a split-screen vision" of nature, which recalls Norgaard's "double reality" and which, Taylor argues, originated in the nineteenth century. It is characterized, in his account, by the sundering of "the scientific explanation of the natural order" from its "moral meaning."[29] His discussion of Wilson's *On Human Nature* occurs in a chapter called "Our Victorian Contemporaries," which traces a number of still-unresolved moral and epistemological fissures in the secular imagination back to their roots in nineteenth-century science. And, as we'll see, many Victorians shared Wilson's interest in the value of "blind hopes" as a means of attending to "the deepest needs of human nature," which might otherwise be threatened by the picture of reality being painted by evolutionary biology. And yet, it seems clear today that the real threat to our "deepest needs" inheres not in confronting our diminished ontological significance but in continuing to avoid facing it. It seems clear, indeed, that our *deepest* needs demand not a mythopoetic construct to help us organize a coherent, developmental narrative of "hope" for our species but rather a livable planet for ourselves and our non-human brethren, which is precisely what our Promethean ambitions have put at risk.

Of course, if other potential organizing myths are being pushed aside, we might wonder what else is getting sidelined for the sake of Wilson's chosen evolutionary epic. Who is included in the "we" being hailed, and, more importantly, who is excluded from it? What groups, exactly, are imagined to enjoy "some measure of dominion over the physical environment"? Wilson's implied answers to these questions are also at heart quintessentially Victorian, even if he takes more rhetorical pains than his nineteenth-century predecessors to whitewash the imperial and racial hierarchies that structure his narrative of progress:

> Man's destiny is to know, if only because societies with knowledge culturally dominate societies that lack it. Luddites and anti-intellectuals do not master the differential equations of thermodynamics or the biochemical cures of illness. They stay in thatched huts and die young. Cultures with unifying goals will learn more rapidly than those that lack them, and an autocatalytic growth of learning will follow because scientific materialism is the only mythology that can manufacture great goals from the sustained pursuit of pure knowledge.[30]

The word "manufacture," literally used to describe the generation of ideas, clearly also has a material ring, suggesting that the mythological *itself* is a source of motive power: "autocatalytic" and self-sustaining. Believing something is real in some ways makes it so. This, as we'll see and as I've written about elsewhere, is also a very Victorian kind of slippage, where energy the organized material resource is confounded with energy the organizing (or "unifying") quality of character or culture in a way that obscures its materiality.[31] But I've buried the lede here, because the much stranger part is the beginning of the passage. Because in what sense, it seems fair to ask, can "anti-intellectuals" be considered a "society"? Certainly not in the sense implied by the first sentence, which seems to be gesturing to a much broader scale—the national or imperial—and the historical questions of which groups of people "dominate" others. The image of "thatched huts" is obviously a loaded one, and it conjures neither Luddites nor anti-intellectuals but rather indigenous societies and inhabitants of the Global South. Wilson can't *quite* say this out loud, but this is clearly a carefully coded Social Darwinist argument for, and rationalization of, Western global hegemony as the great hope for humanity.[32] It is as clear an example as one could want of the dynamic Sylvia Wynter describes as "the overrepresentation of Man," which is the way that "the present ethnoclass (i.e., Western bourgeois) conception of the human, Man . . . overrepresents itself as if it were the human itself."[33] In her account, discussed further later, the newly reimagined, inclusive, biocentric category "human" emerges in evolutionary writing as it gets, at the same time, effaced by the exclusive, hierarchically structured, sociogenic fabrication "Man." Inclusion becomes exclusion through the logic of disavowal—signaled in Wynter's "as if." No one actually believes that the Western bourgeois ethnoclass *is* the human itself; all the same, so much cultural and economic life seems tacitly organized upon precisely this hierarchy that it becomes impossible to conclude this is not, in fact, what many people actually believe. Indeed, the pervasiveness of such an assumption is what allows Wilson to not quite say what he means but know the message is coming across anyway. The point here is that the discourse of progress, and the need to find some way to demarcate the human from the rest of the natural world—whether metaphysically, historically, mythopoetically, or narratively—inevitably depends upon racialized exclusions and the logic of white supremacy. Those exclusions are sometimes explicitly owned, sometimes tacitly assumed, sometimes only ambiguously voiced or acknowledged, but in all cases they make only a privileged subset eligible to be included in the mythmaking.

"How do we hold the two together," Dipesh Chakrabarty asks, "as we think the history of the world since the Enlightenment?"[34] Chakrabarty's question is to some degree an attempt to address the problem of disavowal on the level of the historiographical: that is, to figure out how to integrate competing frameworks for the human, each of which has a claim upon the "real," but each of which establishes discursive boundaries that stabilize one aspect of the object of study while excluding another, equally important, aspect. What to do with "Anthropocene," "species," and "human" when the very universalizing power that makes such concepts valuable in one kind of analytical framework is exactly what obscures the histories through which such universals have been constructed, as Wynter shows, on a hierarchical racial logic of exclusion? To track the history of those constructions is necessarily to track the history of colonialism and empire and the suicidally extractivist drive that produced the so-called Anthropocene in the first place. The philosopher Axelle Karera argues that the unexamined use of such universalizing terms means the social, political, and historical dimensions of the climate crisis routinely get ignored in much environmental and new materialist discourse: "The new regimes of Anthropocenean consciousness have been powerful in disavowing racial antagonisms."[35] Such erasures, she argues, risk replicating the deep structure that generated, and continues to generate, the crisis: "what gets lost when the urgencies of anxious petitions to obtain 'a solution' aggressively compete with the difficult work of finding the conditions that render stable the repetitions and the duration of anti-blackness, or the legacies of slavery and colonialism."[36] Thus, what scholars like Chakrabarty, Wynter, and Karera make plain is that it is no contradiction to find that the Promethean Wilson, who wove his disparaging insinuations around people living in "thatched huts," was also a committed environmentalist and outspoken critic of that reckless "planetary killer," humanity.[37] Chakrabarty notes that Wilson's solution to environmental crisis is more knowledge, more self-awareness, more technological sophistication—"the unity possible through our collective self-recognition as a species," as he glosses it.[38] "Humanity," Wilson writes, "has consumed or transformed enough of Earth's irreplaceable resources to be in better shape than ever before. We are smart enough and now, one hopes, well informed enough to achieve self-understanding as a unified species."[39] The idea that humanity's collective consumption of energy will be transformed into a universal species brain to help us manage the disasters that consumption has caused is a daft fantasy, a story straight out of the nineteenth century and its narratives of inevitable human progress. It also rings of imperial mythmaking and the belief that there exists a special "we" destined to benevolently administer the world, its peoples, its resources. It's a story we'll see repeatedly in the following pages: the promise of the emancipatory power of knowledge that turns out to be just one more way of not knowing.[40]

In an address delivered a few years before Dillard wrote *Pilgrim at Tinker Creek*, Rachel Carson told her audience that "today, it would be hard to find any person of education who would deny the facts of evolution. Yet so many of us deny the obvious corollary: that man is affected by the same environmental influences that control the lives of all the many thousands of other species to which he is related by evolutionary ties."[41] The toxic effects of pesticides and other industrial chemicals are well understood, she says, and yet, incredibly, they still continue to be fed heedlessly into the air, water, and soil "to serve the gods of profit and production."[42] Turning away from the insects in loathing has metastasized, via big agribusiness, into eradicating them wholesale. The disconnect between environmental knowledge and economic practice has everything to do with the incentive structure of capitalism where short-term gain is privileged over long-term health and an illusory sense of human difference is strategically conjured to keep concerns about the latter at bay. Evading the radically species-decentering thrust of Darwin's theory—which we might gloss briefly and incompletely here as the non-hierarchical enmeshment of the human and the non-human—means evading a clear-sighted reckoning with our status as biological beings and thus the vulnerability of our own animal bodies to the chemical warfare we have unleashed upon other animal and plant species in the name of ever-increasing profits. It is a failure to acknowledge the manifold forms of environmental damage being visited at and across nearly every imaginable scale—from the microbiological to the planetary—and through much wider temporal horizons than the quarterly demands of the profit motive can admit. For Carson, the ontological and the ecological are entirely bound together and co-constitutive, which means that the denial of evolution and the denial of environmental destruction are not just structurally similar but are interarticulated aspects of the same phenomenon. The relationship is both analogical and genealogical, and the failure to fully realize the "obvious corollary" of Darwinian thought—that we are fully enmeshed in material environments like all other organisms—is both sign and cause of a deliberately impoverished ecological imagination, a culture of denial.

This book is about that culture: the way it was shaped by (as it helped shape) the response to Darwin's theory in the nineteenth century; its warping effects on environmental thought in that period and beyond; its relationship to other forms of social and political evasion, especially those connected to the imperial project and racialized hierarchies of humanness; and its expression in the literary imaginary of the long nineteenth century. My focus in these pages is primarily on soft rather than hard denial, although these seemingly distinct responses are more difficult to distinguish than first appears and were, indeed, often entangled and interarticulated in the Victorian period, as they are today. As we'll repeatedly discover, the very focus on the stark irrationality of hard denial can help tacitly normalize soft denial and make it seem harmless by comparison. This is partly what Klein is after in her discussion of these terms: the very desire to demarcate (my, understandable, harmless) soft from (their, irrational, destructive) hard denial actually highlights their interconnections.

In recent years, there has been something of a renewed interest in Darwin in the humanities, especially as ecological concerns have become more insistent and central to literary studies, critical theory, anthropology, sociology, history, and any number of other disciplines. Not a *revival* of Darwin, exactly, because that would imply he went somewhere, but something of a new emphasis on, and interest in, Darwin the theorist, the thinker who radically challenged the place of the human in the natural order and, along with it, many of the most basic categories of Enlightenment thought.[43] For the literary critic Timothy Morton, Darwin is a deconstructionist *avant la lettre*, upending not just the binaries *animal* and *human*, but *organism* and *habitat*, and even *subject* and *object*. Such distinctions and reifications need to be broken down, in Morton's view, in order to bring a new kind of environmental imaginary into being, and it is Darwin who begins to show the way, "who thought through many of the complex and hard-to-face issues that confront the ecological thought."[44] In a similarly deconstructive vein, Elizabeth Grosz puts Darwin's reconceptualization of life at the center of her own eco-feminist project and argues that "philosophy has yet to recover from this eruption, has yet to recompose its concepts of man, reason, and consciousness to accommodate the Darwinian explosion."[45] Devin Griffiths, meanwhile, argues for understanding Darwin "as a foundational process philosopher and ecotheorist" and persuasively argues that "Darwin's theory of natural selection has overshadowed his more audacious lifelong project: to explain how, in the absence of a design and in the face of constant flux, natural patterns sustain themselves."[46] Darwin's ecology, Griffiths argues, provides us with a resource for thoroughly reimagining our understanding of form, literary and otherwise. We can also see the interest in Darwin the theorist in recent work by Donna Haraway and Bruno Latour, as well as work in the so-called animal turn in the humanities.[47]

This book, you might say, half-participates in this critical trend. That is, it is also interested in these radical, bracing, paradigm-exploding, eco-theoretical aspects of Darwin's work, but it is just as focused on the ways in which such challenging ideas were avoided, downplayed, euphemized, side-stepped, twisted, evaded, and denied (and continue to be). You can see this tension in the brief quotes earlier by Morton and Grosz: the implications of Darwin's ecological ideas are "hard to face," difficult to "accommodate," and this book explores not only all of the ways the Victorians avoided facing and accommodating them but also the many ways in which that avoidant response persists. Evolutionary science's decentering of the human was so extreme—Herbert Spencer, as we'll see, called it a "laceration"—that it seemed to have encouraged a powerful, compensatory recentering project on other, more contingent, precarious, and in some cases knowingly fictitious grounds. Before Darwin, the idea of human exceptionalism always had the immortal soul to fall back upon; this meant that denial involved the refusal to accept the animal *part* of the human rather than the refusal to accept the human *as* an animal. But the human-as-animal is an immensely more challenging proposition because it punctures many cherished and deeply rooted beliefs, explicit and implicit, about the possibility of human transcendence over the merely material and the role some humans want to give themselves as the special avatars uniquely equipped to realize that possibility. It's no wonder, then, that so many Victorians struggled to face it.

In what follows, I discuss the ways in which this struggle expressed itself in the literature of the nineteenth century and beyond. In Chapter 1, I examine the concept of denial, discussing its connections to Marxist ideology critique; its operation in the history of imperial conquest and white supremacy; and its development in Freudian psychoanalysis. One of the key aims of this chapter is to show the complex interdependence of various forms of denialist thinking and how they might be understood in relation to the challenge to human exceptionalism posed by Darwinian theory. Ian Duncan describes how late eighteenth- and early nineteenth-century ideas about human development made a case for human exceptionalism that was ultimately self-undermining: "The formal qualities supposed by late Enlightenment philosophers to set man apart in nature—plasticity, perfectibility—embed him more deeply within it."[48] In Chapter 2, I'm interested in something like the other side of this coin: not in the upending of human exceptionalism through the science meant to fortify it but in its ironic reconsolidation through the science that had seemingly dismantled it. That is, the way that, after Darwin had systematically and decisively undermined anthropocentric thought, nineteenth-century evolutionary scientists and anthropologists used a version of his ideas to provisionally return "humanity" to a place of centrality in the natural order. The features of human existence that Darwin made increasingly difficult for secular humanists to flatly deny—our animal nature; our full enmeshment in natural systems and processes; our inevitable extinction—could still be "softly" denied: pushed out of mind, temporarily suspended, explained away. This took place through a developmentalist and indeed "Promethean" mythmaking that provided cover for the imperial project, as that project helped organize the entire discursive field of scientific inquiry. Crucially, this meant that the radical *ecological* import of *The Origin of Species* was evaded as soon as it was articulated, resulting in a divided view of the natural environment in which a select minority of humans were somehow imagined to be *from*, but not *of*, nature.

This book is thus interested not so much in the ways literature refuses, suppresses, or excludes Darwinian ecological ideas but in how, on what terms, to what extent, and with what implications (for morality, affect, politics, form) it *admitted* them into a given field of representation. Where are the borders drawn, how are principles of inclusion and exclusion defined, how does all that is going unacknowledged get acknowledged, and how does acknowledgment itself sometimes function as a dodge? Organizing the discussion of Victorian and modernist texts in Chapters 3, 4, and 5 is the question of "realism," but only if that term is understood in a capacious and somewhat unorthodox sense. That is, I discuss those supreme monuments of realistic fiction, George Eliot's *Middlemarch* and *Daniel Deronda*, and the ways they give vivid expression to the operations of disavowal. But I also consider novels like *The Time Machine, Heart of Darkness, Dr. Jekyll and Mr. Hyde, The Island of Dr. Moreau*, and *To the Lighthouse*, among others, not to convince you that these should all be reclassified as "realist" but to show how they all stage, in various ways, through various literary modes and formal devices, a cognitive and affective clash between the ordinary world that is the province of realist representation and the more disorienting and destabilizing evolutionary "realities" described by Darwinian science. What unites this array of texts is a common interest in the construction and reconstruction of, the departure from and the return to, the ordinary, human-scaled, "real" world. Chapters 3 and 4 are further organized around the question of voice: the former focusing on the question of self-consciousness in first-person narration and how the felt experience of inner autonomy and personal significance sits uneasily with a conceptual understanding of the profound cosmic decentering of both individual and species entailed by Darwin's theory. This contradiction is further complicated by the fact that the capacity for self-conscious reflection was frequently claimed as the final line of demarcation between fully human and other, non- or only partly human, existences. This could take the form either of a tautology—*we know we're different because we know we're different*—or a paradox—*we know we're different because we know we're not different*. Disavowal was thus often encoded in the very formal structure of first-person texts, dwelling in an experience of distinction they also know is illusory.

Chapter 4 pivots from first to third person to consider the "dual voice" of free indirect discourse and the way the merger of the "omniscient" narrator with the limited, self-dealing, but also always potentially self-aware perspective of a given character, creates a necessarily indeterminate zone of knowing and unknowing, making this technique uniquely suited to expressing the divided state of mind known as denial. Beginning with a discussion of free indirect discourse in a number of nineteenth-century writers, the chapter then turns its attention to Eliot's *Middlemarch* and *Daniel Deronda* and the way those novels use it to both voice and muffle the destabilizing implications of Darwinian science. Chapter 5 focuses on Virginia Woolf and her masterful exposure of Victorian denialism in *To the Lighthouse*. I consider how her novel registers the allures of the organizing consolations of bourgeois life while dramatically, breathtakingly, extending the field of representation to admit the inhuman scales and realities that constantly press in upon the awareness of her characters and the stories they tell themselves. Woolf draws upon her lifelong interest in Darwin, and her preoccupation with the culture of denial bequeathed by her Victorian predecessors, to directly connect the evasion of natural realities to the forms of willful blindness that resulted in the Great War. By making vivid the relationship between widespread denialism and an otherwise avoidable, human-made, global catastrophe, Woolf's novel looks both forward and back, anatomizing the Victorian past while imagining the ways it continues to inflict itself upon the environments of the future. The book ends with a brief(ish) concluding section that features a surprise guest whose books helped establish much of the conceptual groundwork for this one. The Conclusion considers the persistence, in contemporary works of fiction, of the questions I track through the nineteenth century, including how the formal and aesthetic resources of the novel continue to be employed to express the unresolved fissures in the secular liberal imagination of Darwin and the implications for environmental thinking. The challenge Darwinian theory poses to human stories, the human image, the construction of "everyday" reality can be said to contranymically "cleave" all of the texts I discuss, creating splits and fractures that also attach and bind. Disavowal, as we'll see, is form making and form breaking at once.

"How do we hold the two together?" Chakrabarty asks. It's a question Darwin and his fellow scientists ask about the conflicting images of the human coming into being in the wake of his theory. It's a version of the question George Eliot asks about the place of science in the liberal imaginary, how Darwin's work could at once open up new possibilities for humanist reimaginings of social organization while also threatening to destroy the very boundaries and structuring assumptions needed for that reimagination to occur. It's a version of the question Woolf poses to her readers when she insists upon the absolute significance of the daily life of the Ramsay family and its sphere of human meaning and its absolute insignificance against the backdrop of the ocean, the elements, the catastrophic convulsions of natural and human history. This book is all about the ways this question gets asked—the forms it takes, the terms in which it is put, the pressures it exerts on perspective, structure, and voice—as well as the many ways it gets answered and evaded, approached and dodged, in the never-ending story of the nineteenth century.

ONE

FORMS OF DENIAL

IT IS "HARD" SCIENCE DENIAL that usually attracts our attention, and for good reason. So-called climate skeptics are creatures of the corporate and religious right, bankrolled by moneyed interests, cynically raising doubts and hawking delusions in the service of their reactionary agenda. As Andreas Malm has argued, climate science, and everything it tells us about the urgent need to decarbonize the entire global economy, represents an existential threat to the oil and gas industry, or what he calls "primitive fossil capital," which must eventually be shut down altogether if there is any chance of keeping the planet habitable.[1] In its desperate attempt to maintain its own existence, fossil capital has created what he calls a "denial machine" of think tanks and news outlets and other devices now set in open war with life itself. But the entire capitalist order is threatened by climate science—structurally, if not always existentially—and industries more indirectly tied to fossil-fuel extraction (the automotive or steel industries, say, which, as Malm notes, could at least theoretically survive wholesale decarbonization) have shown themselves more than willing to make their own contributions to the denial machine.

The hard denial of science was a different animal in the nineteenth century. For one thing, it was less cynically organized and strategically interventionist in the public discourse. In part, that's because the science itself was much less settled and much less obviously or directly in conflict with capitalism—indeed, some interpretations of it seemed very friendly to a capitalist worldview, as Marx, among others, famously noted.[2] It was also the case that spiritual belief in the nineteenth century was a more ubiquitous and therefore more redoubtable cultural bulwark against materialism. Science denial wasn't bankrolled by the same kind of malign, deep-pocketed interests, not because there weren't malign, deep-pocketed interests to bankroll things in the nineteenth century but because it didn't need to be. Nevertheless, there are continuities—we might even say family resemblances—in the strategies used to sow doubt, then and now. So overwhelming, for example, was the evidence for the vast age of the earth, the mutability of species, and other foundational components of the evolutionary paradigm that some opponents pushed back by undermining the very idea of evidence itself. Such was the strategy notoriously taken to extreme lengths by the naturalist Phillip Gosse, who argued in *Omphalos* that, okay yes, the geological record seems to indicate the earth is immeasurably older than the Bible says, but that's just because God planted fake fossils of extinct species in the ground to make a young earth *look* old, to test people's faith.[3] "We believe the logic of the book to be unanswerable," a contemporary reviewer in the *Natural History Review* snorted, and indeed, Gosse's mode of argument works by turning doubt into the very basis for belief while rendering empirical inquiry essentially meaningless.[4] Such an impervious epistemological bubble recalls the logic of the Queen's court in *Alice in Wonderland* and looks ahead to the white board and strings of contemporary conspiracy theory; in this case, no lesser personage than God himself is out burying false flags in the shale. This is denial in its starkest and most risible form—akin to, say, bringing a snowball onto the Senate floor to disprove climate change—and, in this sense, Gosse's work is an outlier. But the shapes into which it contorts itself to avoid the obvious also make it a quintessential expression of the desperation many felt in the face of this supreme challenge to the Christian order. For the Victorian evolutionists, such a book was as clear a demonstration as there could be of the lengths to which their opponents would go to avoid facing the obvious.

Gosse's strategy—interpret the evidence clean out of existence—was unpersuasive. More effective was what the Swedish philosopher Sven Ove Hansson calls the application of "deviant criteria of assent"—the demand for an evidentiary standard so high as to put proof forever out of reach—which, he argues, is a strategy common to both climate denial and the denial of evolution.[5] In a review of Darwin's *Origin*, Bishop Samuel Wilberforce (he of Huxley-Wilberforce debate fame) demanded to know why we don't see examples of plants turning into animals today: "The closest microscopic observation has never detected the faintest tendency in the highest of the Algae to improve into the very lowest Zoophyte," he says, so there is no reason to believe, along with the evolutionists, that "favourable varieties of turnips are tending to become men."[6] Darwin wrote "rubbish!" next to this passage in his copy of the review, but such an approach—an impossible demand for tangible proof (coupled, in this case, with a misleading account of the theory)—is on page 1 in the playbook for creationists then, as well as for flat-earthers, so-called climate skeptics, and any number of other "merchants of doubt" today.[7]

It would be a vast oversimplification to categorize all Victorian objections to Darwin as forms of "denial," because that would be to ignore the many, complex, very real, unresolved questions that attended his theory in the later decades of the nineteenth century and beyond. But the *desire* to avoid the unsettling implications of an entirely chance-driven, directionless, and inhuman natural order also helped make refutations and alternatives especially appealing, and many writers were quite candid about the nature of that appeal. The boundary between what one thinks and what one *wants* to think is far from clear, and the Victorian conversation about evolution brought this tension to the surface, both as rhetorical strategy and as explicit topic of consideration. In an early discussion of *The Origin of Species* in *Blackwood's*, the anonymous author admits feeling reassured to learn about what he considers serious flaws in Darwin's model—in this case, the oft-leveled (if decidedly wrongheaded) critique that natural selection cannot account for the persistence of "lower" organisms into the present:

> We confess that we have been relieved to find that we are not expected to consider man as merely an improved mollusk—highly as we estimate many points in the character of those animals. Mr. Darwin's theory, so far as we understood it, was, to say the least, very uncomfortable. That man has a good deal of the beast about him, we admit with shame; but to adopt every beast in creation into our own family tree is a more comprehensive genealogy than we can bring our minds to.[8]

Such a "confession," archly phrased though it is, is actually rather startling: the main problem with *The Origin of Species* is that its implications are uncomfortable and disconcerting, simply too difficult to face. Good thing it isn't true, then! We will see in the next chapter how, in *The Descent of Man*, Darwin confronts these questions of shame and admission directly in an attempt to help his reader get past such aversive reactions.

Although the nineteenth century didn't have the same "denial machine" of semi-coordinated think tanks and corporate media outlets and lobbyists and fake scientists and bot farms disseminating misinformation, it did have its own powerful institutional networks of opposition. Here, a critic in the *Dublin Review* speaks on behalf of one of them: "The salvation of man is a far higher object than the progress of science: and we have no hesitation in maintaining that if in the judgment of the Church the promulgation of any scientific truth was more likely to hinder man's salvation than to promote it, she would not only be justified in her efforts to suppress it, but it would be her bounden duty to do her utmost to suppress it."[9] A remarkable statement, given how spectacularly wrong the Catholic Church had been in its crackdown on, say, Galileo's work, to use only the most notorious example. But the writer tries to turn this infamous history to his advantage: "The truth ultimately can do no harm, although, temporarily, injury may follow from an unseasonable application of it."[10] The church doesn't stop us from arriving at the truth, but it does tap the brakes to keep us from crashing. This was not an uncommon way to challenge Darwinian theory in the nineteenth century—leave the science alone and play up the dreadful moral and social implications. Of course, such an approach all but gives the game away insofar as it amounts to a confession of motivated reasoning. Consider, as another example, this moment from the pages of the *Edinburgh Review*:

> It is indeed impossible to over-estimate the magnitude of the issue. If our humanity be merely the natural product of the modified faculties of brutes, most earnest-minded men will be compelled to give up those motives by which they have attempted to live noble and virtuous lives, as founded on a mistake; our moral sense will turn out to be a mere developed instinct . . . and the revelation of God to us, and the hope of a future life, pleasurable daydreams invented for the good of society. If these views be true, a revolu-

tion in thought is imminent, which will shake society to its very foundations, by destroying the sanctity of the conscience and the religious sense.[11]

If social media had been around in 1871, this writer's mentions would be filled with "he's so close to getting it . . ." Other opponents of Darwin went after the science itself, some posing legitimate unanswered questions, others wielding questions *as if* they were unanswerable. The question in all of these cases is not whether this or that writer was "in denial," whether they were sincere in their opposition, consciously acting in bad faith, or dwelling in some ambiguous, shifting zone of mixed feelings, attitudes, and pressures. Whatever their motivations or psychological states, they helped construct a cultural discourse of hard science denialism that worked to sow widespread doubt about Darwin's theory in the Victorian public sphere.[12]

And doubts are easy to sow, then and now, because evolution, like climate change, is what Timothy Morton calls a "hyperobject," a phenomenon so "massively distributed in time and space" that it exceeds the capacities of the human sensorium: "Stop the tape of evolution anywhere and you won't see it. Stand under a rain cloud and it's not global warming you'll feel."[13] As Darwin himself puts it, "The mind cannot possibly grasp the full meaning of the term of a hundred million years; it cannot add up and perceive the full effects of many slight variations, accumulated during an almost infinite number of generations."[14] This imaginative gap is therefore a perfect place for opponents like Wilberforce to set up shop: *Hey, I stopped the tape and I didn't see anything!* Hyperobjects are real, but unavailable to direct experience; they can be known only through abstract thinking, but they are not themselves abstractions. Such a paradoxical situation, as George Levine says, "came increasingly to occupy the attention and imagination of nineteenth-century writers."[15] Indeed, you can see Darwin attempting to find language for it throughout *The Origin of Species*, as in this famous passage:

> It may be said that natural selection is daily and hourly scrutinising, throughout the world, every variation, even the slightest; rejecting that which is bad, preserving and adding up all that is good; silently and insensibly working, whenever and wherever opportunity offers, at the improvement of each organic being in relation to its organic and inorganic conditions of life. We see nothing of these slow changes in progress, until the hand of time has

marked the long lapse of ages, and then so imperfect is our view into long past geological ages, that we only see that the forms of life are now different from what they formerly were.[16]

The notion that natural selection works "silently" is counterintuitive, considering how central violence is to its operations. We might recall that what follows Alfred Tennyson's startlingly visual line "Nature, red in tooth and claw" is an ear-splitting screech: "With ravine, shriek'd against his creed."[17] The evolutionary motor runs on blood, but Darwin emphasizes silence because the growls and screams of predators and prey are just the audible manifestations of the workings of natural selection, not the thing itself. This is not to say that those material phenomena are any less "real" than the theories used to make sense of them; in fact, as Morton argues, "Hyperobjects compel us to think ecologically and not the other way around. It's not as if some abstract environmental system made us think like this; rather, plutonium, global warming, pollution, and so on, gave rise to ecological thinking. To think otherwise is to confuse the map with the territory. . . . Hyperobjects are not the data: they are hyperobjects."[18]

But because the operations of evolutionary (and geological, and climatic) change occur on scales far beyond the bounds of human perception, it takes some imaginative effort—along with the willingness to make it—to grasp them. Not everyone feels compelled to put in that effort, and even fewer are willing to bear it "constantly" in mind. And that, in turn, makes hyperobjects ripe for misunderstanding, whether willful or otherwise. As Levine writes, "Ironically, in a literature largely committed to an empirical understanding of the world, science was replacing the unseen reality of a world of spirit with another unseen reality, the world of matter—a world unheard or barely heard that is more active, more alive, than the one we can hear and see."[19] What Levine describes as an irony can be easily made to seem like a contradiction, a betrayal of first principles, which is why Wilberforce triumphantly brandishes his microscope. The good bishop would have you think he has traded places with the scientists, taking up the cause of rigorous empirical inquiry they seem to have abandoned.

The difficulty imagining evolution is not just cognitive but affective, and, as we saw with the critics mentioned earlier, Darwin's theory raises questions not just about what people *can* think but what they *want* to think about. In *The Lifted Veil*, a proto-science-fiction novel that appeared in the same year as *The Origin of Species*, George Eliot considers the positive value of leaving some veils unlifted:

> So absolute is our soul's need of something hidden and uncertain for the maintenance of that doubt and hope and effort which are the breath of its life, that if the whole future were laid bare to us beyond to-day, the interest of all mankind would be bent on the hours that lie between.... Conceive the condition of the human mind if all propositions whatsoever were self-evident except one, which was to become self-evident at the close of a summer's day, but in the meantime might be the subject of question, of hypothesis, of debate. Art and philosophy, literature and science, would fasten like bees on that one proposition which had the honey of probability in it, and be the more eager because their enjoyment would end with sunset. Our impulses, our spiritual activities, no more adjust themselves to the idea of their future nullity, than the beating of our heart, or the irritability of our muscles.[20]

The plot of *The Lifted Veil* revolves around a character with ESP who can see the future, often down to the smallest detail. But this passage, in its concern for "the whole future," the fate of life in its entirety, also voices an anxiety about human extinction, a prospect much in the air at this moment. Eliot here expertly plays with the paradoxes of the human-as-animal, implying that it is our throbbing, irritably muscled, creaturely life that resists the morbid, abstract forecast of our "future nullity," our ultimate fate *as* a mere animal species. Meanwhile, and even more counterintuitively, our "higher" cultural activities are themselves, deep down, mere expressions of instincts and appetites—hive-like rituals—even if those activities function to distract us from that very proposition. Without such distractions, Eliot suggests, all of human experience appears as little more than a compost pile of pointless and undifferentiated matter-energy: "The rational talk, the graceful attentions, the wittily-turned phrases, and the kindly deeds, which used to make the web of their characters, were seen as if thrust asunder by a microscopic vision, that showed all the intermediate frivolities, all the suppressed egoism, all the struggling chaos of puerilities, meanness, vague capricious memories, and indolent makeshift thoughts, from which human words and deeds emerge like leaflets covering a fermenting heap."[21] The microscopic leads inevitably to the macroscopic, just as Darwin's work takes us from the drives of survival and reproduction that impel an individual organism to wide-angle views of population dynamics, gradualist models of change, and the inevitability of extinction. Eliot's novella runs along the divide we'll be tracking in this book—between, on the one hand, knowledge about the material systems in which all human existence is embedded, including the kinds of scalar immensities that reduce that existence to a mere speck, and, on the other, the various cultural endeavors that are structured by—as they also structure—the desire not to know our position or ultimate end point. The make-believe of a middle. The despair-inducing implications of scientific knowledge are thereby fended off by means of a complex trick, as *The Lifted Veil* slyly suggests that the true purpose of culture might be to conceal the true purpose of culture, which is to conceal.

And that, Eliot thinks, might not be such a bad thing. "Of scientific truth," she wrote in her posthumously published "Leaves from a Notebook," "is it not conceivable that some facts as to the tendency of things affecting the final destination of the race might be more hurtful when they had entered into the human consciousness than they would have been if they had remained purely external in their activity?"[22] Note that, by setting up a division between inside and out, between what is given permission to "enter" and what must remain outside, Eliot, like Darwin, plays on the metaphorics of "admission." In considering whether it might be too painful to the "human consciousness" to handle the reality of extinction, she seems to be imagining that prospect both in terms of the distress experienced by the individual subject and a broader kind of injury inflicted upon a social order plunged into despair and relativism. There is, in other words, more than a hint here of the concerns famously voiced by Dmitri Karamazov: "How will man be after that? . . . Without God and the future life? It means everything is permitted now, one can do anything?"[23] Eliot's liberal humanism is a far cry from Dostoevsky's Christian conservativism, but this is precisely the point—the soft denial of Darwin on the part of secular humanists is not as clearly distinguishable from the hard denialism of religious believers as one might think. Indeed, Eliot's misgivings about the wider implications of certain aspects of Darwinian thought are not unlike those expressed by the anonymous critics in the *Edinburgh Review* and the *Dublin Review* quoted earlier: whether certain things are true or not, maybe we would just be better off not knowing them. Eliot is less intransigent than her religious-minded counterparts, and her denialism is voiced through a denser rhetorical thicket of qualification and hesitation. This is, in part, because willful ignorance sits uneasily with her entire moral, social, and epistemological project, as we'll see in more detail in Chapter 4. Evolutionary science was something of a double-edged instrument in secular liberal thought: although it provided a powerful set of arguments against a Christian cosmogony that formed the backbone of the conservative political order, it also (less obviously, but no less deeply) struck at many core liberal assumptions about progress, rationality, moral agency, and human preeminence. Eliot, in other words, has good reason to accept and reject Darwin at once.

Eliot's chosen trope, the veil, was commonly used in the period as a figure for soft denial or disavowal, especially where scientific subjects were concerned. Thomas Huxley, for example, invokes it to complain about what he considers the willful blindness of the geologist James Hutton: "He persistently, in practice, shut his eyes to the existence of that prior and different state of things which, in theory, he admitted; and in this aversion to look beyond the veil of stratified rocks, Lyell followed him."[24] A veil both conceals and beckons; it obscures the hidden thing while drawing attention to the hiding place; it is easy to remove, and thus its placement is always wrapped up in questions of will and desire. Meanwhile, the precise nature and quality and extent of the concealment are always left ambiguous—veils can be diaphanous, verging on transparent; or they can shroud in opacity. The trope is thus evocative but difficult to picture and usefully indistinct; the veil, in a sense, is veiled. This allows it to figure something of the forever-unresolved both/and quality of disavowal—the uncertainty it conveys about how much one can afford to know or not know, and the formal gamesmanship involved in the self-conscious participation in make-believe.

We can see such gamesmanship in an even more famous use of the trope in the period, from Tennyson's *In Memoriam*. Just after the famous "Nature, red in tooth and claw" moment, the speaker expresses his despair about the futility of life in the face of geological time. His solution is to find relief in unknowing: "What hope of answer or redress? / Behind the veil, behind the veil."[25] At first glance, it would seem that Tennyson's use of the trope is something like the opposite of Eliot's—hope, rather than horror, awaits us once the veil is pulled back. But the whole point is that such hope can be posited only because the veil is *not* pulled back; if one keeps it in place, if one resists accepting the kind of relentless "unveiling" of material reality being performed by geologists and evolutionists, then one can go back to putting whatever one wants on the other side. Note how the repetition of the phrase "behind the veil" not only gives the line an incantatory quality; it also inflects the preposition with a hint of the imperative. The road to consolation is not to look behind the veil but to make sure you keep things there. The ambiguity in the trope is compounded by the spatial ambiguity in the relationship between perceiving subject and unperceived object: to posit "hope" on the other side is to create it on *this* side through the very act of obscuring whatever is on the other side. As Franco Moretti argues, the veil is a not only a figure Tennyson strategically employs in the poem; it is also, in a sense, a way of understanding the very operations of the poem's rhetoric. Discussing the syntactical complexities of the cantos on geology and extinction, Moretti writes: "Poetic intelligence sees mankind's extinction—and buries it within an unfathomable linguistic maze. . . . Take the truth that has somehow emerged and place it in brackets."[26] Doubleness is encoded in the language itself—concealment occurring through the very act of expression. Gillian Beer hears a similar dynamic at play in Eliot's discussions of natural selection:

> [It is] an idea so much at variance with George Eliot's own morality that it is not surprising that she did not immediately grasp its implications. Whenever she refers directly to the idea of natural selection, that faintly facetious orotund style appears, to which she is driven by ideas that cause her deep disquiet and which she yet cannot repudiate.[27]

Cannot quite accept and yet cannot repudiate—the disavowal of natural selection, the most essential *and* the most radical part of Darwin's theory, creates warps and fissures that become audible in the prose. For Klein, the "soft" in the term "soft denial" indicates a metaphorical logic of flexibility, but we might also understand it as a question of voice. Soft denial is denial spoken sotto voce, under the breath, in equivocal rhetorical arrangements and through verbal channels that stifle and dampen even as they circulate and express.

One further example, this time from Charlotte Brontë, who uses the same trope to ward off the same "horror" of materialism and extinction.[28] In a letter to James Taylor, she describes her response to Harriet Martineau's *Letters on the Laws of Man's Nature and Development*, a book that preceded *The Origin of Species* by about nine years but was part of a widespread speculative engagement with evolutionary matters that set the stage for Darwin's work:

> Of the impression this book has made on me, I will not now say much. It is the first exposition of avowed atheism and materialism I have ever read; the first unequivocal declaration of disbelief in the existence of a God or a Future Life I have ever seen. In judging of such exposition and declaration, one would wish entirely to put aside the sort of instinctive horror they awaken, and to consider them in an impartial spirit and collected mood. This I find it difficult to do. The strangest thing is that we are called on to rejoice over this hopeless blank—to receive this bitter bereavement as great gain—to welcome this unutterable desolation as a state of pleasant freedom. Who <u>could</u> do this if he would? Who <u>would</u> do it if he could? Sincerely—for my own part—do I wish to know and find the Truth—but if <u>this</u> be Truth—well may she guard herself with mysteries, and cover herself with a veil. If this be Truth—man or woman who beholds her can but curse the day he or she was born. I said, however, I would not dwell on what I thought; I wish rather to hear what some other person thinks; some one whose feelings are unapt to bias his judgment. Read the book then in an unprejudiced spirit and candidly say what you think of it. I mean—of course—if you have time—<u>not otherwise</u>.[29]

Brontë here isn't so much responding to the book itself as responding to her response to it: self-consciously arranging her own relationship to unsettling ideas about the natural world and the human place within it. This too is not what we usually mean when we talk about "science denial," in large part because of how remarkably forthright Brontë is about how she lets desire and aversion shape what she allows herself to know. She is refreshingly honest in confessing to self-deception. That sense of doubleness runs through every aspect of the passage: recommending the book to Taylor while warning him not to read it; discussing the ideas by discussing her refusal to discuss them; trying not to bias him while telling him how to read. The self-awareness of her own doublespeak makes it almost triplespeak. Tennyson's use of the veil has an emotionally self-protective quality: it introduces a rhetorical ambiguity that doesn't just reflect but gives form to and thus makes possible a reassuring uncertainty and indefinite postponement of the matter. For Brontë, likewise, there seems no upside to accepting Martineau's ideas: the only outcome she can imagine is paralysis and despair. Ambivalent though she may be about her own stance, the one thing she seems certain of is that any arguments about the radical emancipatory potential of materialism are entirely unconvincing. Her deferral takes the form of passing the "curse" of the book along, *Ringu*-style, to her correspondent: here, *you* read it. The key to all of this is the way she singles out Martineau's "unequivocal" rhetorical posture, her "avowed" exposition of materialism, as if the real problem is with the single-voiced utterance that allows no room for hedging, or deferral, or alternative possibilities to coexist. What Brontë wants, in other words is (to borrow a phrase from Ali Smith) "to be both": to write in such a way that recommendation and rejection, curiosity and closed-mindedness, honesty and deception can be spoken simultaneously in open and unresolved tones. Brontë wants to be the person who puts the veil back where it was so she can also be the person who forgets what's behind it.

SLANTED AND ENCHANTED

For many, if not most, the response to Darwin's ideas involved some kind of disavowal—that is, an intellectual acceptance of the notion that *Homo sapiens* is simply an advanced primate, one more variation of animal life among many others, accompanied by some kind of emotional, moral, psychological, or intellectual investment in the idea that humans—*some* humans—were nevertheless still set apart, were still imbued with special destiny or significance. As the mathematician William Kingdon Clifford put it, touching both extremes at once, "The creature of clay which you despise is the Lord of Nature and the Measure of all things."[30] That investment in continued human superiority took many forms, but common to many was a reimagining of our exceptional status not as ontologically given but as historically, culturally, and technologically produced. As Tennyson puts it in his late poem "The Making of Man,"

> Where is one that, born of woman, altogether can escape
> From the lower world within him, moods of tiger, or of ape?
> Man as yet is being made, and ere the crowning Age of ages,
> Shall not aeon after aeon pass and touch him into shape?[31]

One could hardly ask for a clearer example of Carolyn Merchant's argument in *The Death of Nature* about the way that the order of civilization, coded as masculine, was imagined to emerge out of, and distinguish itself from, a feminized natural world of emotional and material chaos,[32] or of Sylvia Wynter's argument about Western "Man" simultaneously stepping apart from—and in that stepping apart, claiming to represent—all of humanity.[33] The poem encapsulates the way evolution-as-progress seemed to offer a way to rebuild the human self-image that was destroyed by evolution as merely "descent with modification." Of course, it now also had to be acknowledged that, however "advanced" it might be, human civilization was itself still part of nature and thus as much product and manifestation of the evolutionary process as bulwark against it. Thus a strange picture of that civilization emerges, in which it appears from, but not of, nature: inside and outside at the same time.

And of course certain civilizations were imagined as more outside it than others. The West, it was held, through its various putative advances in science, industry, law, government, philosophy, technology, and the like, made possible a much greater degree of separation from, and control over, the raw forces of the natural world. Indigenous civilizations, in contrast, were imagined not only to be at the mercy of those forces, lacking the wherewithal to manage, exploit, and insulate themselves from their environments; they were also commonly described as captive to their own creaturely instincts and appetites. To fully accept human animality, therefore, seemed to many tantamount to surrendering to the chaos and violence of Darwinian nature, or to the "baser" motives. The construction of nature was thus deeply racially coded, in explicit and implicit ways. This is a familiar story, but it is worth emphasizing that science denialism, rooted in the denial of human animality, the dream for transcendence of or escape from the merely material, and all the other forms of magical thinking that theorists like Becker and Tuan describe, is entirely inseparable from the expansionary drive of the imperial project and the ideology of racial difference. And by "inseparable," I mean that the relationship is mutually constitutive: circular and self-reinforcing. Insisting upon factitious distinctions between groups of humans makes it possible to achieve a sense of greater existential distance from the loathsome reality of the animal; or, flipping the Möbius strip around, insisting upon factitious distinctions between humans and animals makes it possible to achieve a sense of greater distance from the human groups you wish to dominate. Empire makes possible the reinscription of a sense of hierarchy that was threatened by Darwin; meanwhile, the sense of hierarchy already inscribed in the practices and ideological apparatus of empire was there to neutralize whatever radically egalitarian or ecological import Darwin's theories might have held. Indeed, those theories were quickly assimilated in the service of empire building and racial thinking, sometimes abetted (especially in *The Descent of Man*) by Darwin himself. Although, after Darwin, it may no longer have been possible to imagine humanity as the cynosure of all creation in some ultimate, ontological sense, it nevertheless *was* still possible to retain the feeling of centrality for some in a hedged, strategic, provisional, imperially motivated, white supremacist, disavowing sense.

So for all the ecological insight and deconstructive power we may now find in Darwin's work, it was also used to argue for an unapologetically instrumental, ruthlessly extractive, and overtly hostile attitude toward the natural world. Gregory Bateson argues that part of the reason for this rests in the emphasis Darwin placed on *struggle* and how he defined the operative agents in that struggle:

> Darwin proposed a theory of natural selection and evolution in which the unit of survival was either the family line or the species or subspecies or something of the sort. But today it is quite obvious that this is not the unit of survival in the real biological world. The unit of survival is *organism* plus *environment*. We are learning by bitter experience that the organism which destroys its environment destroys itself . . . when you make the epistemological error of choosing the wrong unit: you end up with the species versus the other species around it or versus the environment in which it operates. Man against nature.[34]

Indeed, one thing we'll track in the following pages is the way in which the oppositional binary "man against nature" keeps stubbornly, perversely, arising out of the very theory that made nonsense out of such categories and divisions. In this sense, this book follows in the footsteps of Sylvia Wynter, who explores what she calls the "contradiction at the heart of the Darwinian Revolution," by which she means the split between, on the one hand, the new, fully secular, biologically egalitarian concept of the human he theorized in *The Descent of Man*, and, on the other hand, the "ethno-biological beliefs in the genetic inferiority of nonwhite natives," which of course he also voices in that book and which were pressed into imperial service.[35] The way Darwinian theory was weaponized is summed up, Wynter writes, in the "extermination credo" used by the Nazis: "life unworthy of life."[36] That ghastly phrase makes palpable the impacted starkness of the contradiction, which Frantz Fanon famously captured with similar devastating brevity in his indictment of European hypocrisy: "They are never done talking of Man, yet murder men everywhere they find them."[37] This is the cognitive dissonance, the engine of denial, at the very core of Western liberalism.

For the anthropologist Elizabeth Povinelli, this profound contradiction in the construction of violence in the Western cultural imaginary is produced through what she calls "the disavowal function of the liberal horizon," the always-to-be-bridged, and thus never-to-be-bridged, "opening between liberal promise and liberal actualities."[38] In Povinelli's account, any breach of liberal humanist ideals, including any and all forms of state violence (imperial, racial, carceral, environmental), can be openly acknowledged as a problem still to be overcome—can even be acknowledged as a problem *in* the system and the history of its operation that must be addressed—because of the way the logic of deferral "naturalizes systemic social harm in the social tense of the coming good."[39] Acknowledgment of violence thus perversely functions as a means of denying it, strategically obscuring the ways in which it exists as a constitutive part of the system. This, as we'll see in detail in Chapter 2, is key to the liberal uptake of Darwinian evolutionary ideas that were used to naturalize an ideology of developmental meliorism in which violence today can always be rationalized by reference to the horizon from which the greater developmental good always beckons. This "slanted" version of Darwin (to borrow a phrase from Stephen Jay Gould) reimposed a developmentalist narrative patterning on reality that, at times, Darwin's own rhetoric sometimes seems to encourage and even amplify.[40] This, despite the fact that the linchpin *of Darwin's entire project*, the thing that decisively set it apart from previous theories of evolutionary change, was its refusal of just this assumption of progressive directionality, or telos, in nature.

Darwin's theory is commonly labeled as "disenchanting" for the way it stripped the natural world of any trappings of the sacred; as the philosopher Akeel Bilgrami argues, this process of the desacralization of nature, which has its roots in late seventeenth- and early eighteenth-century science, provided cover for the wholesale exploitation of the natural world under liberal capitalism. He writes:

> If God was not everywhere in nature, a sustained metaphysical obstacle to viewing nature as a source of extraction in the form of deforestation, mining, and plantation agriculture (what we today call agribusiness) was removed. . . . With no metaphysical obstacles remaining, the scale of taking from nature's bounty could be pursued with unthinking and unconstrained zeal. Nature, being brute, could not make demands or put constraints on us. Because it was brute, we did not need to respond to it on *its* terms. All the term-making came from *us*, and the terms we would summon were increasingly the terms of utility and gain, converting the very idea of nature—without remainder—to the idea of natural *resources*.[41]

On some level, Darwin's work can be seen as the culminating point in the process Bilgrami describes. His thorough dismantling of natural theology, of the notion of an anthropocentric cosmic order, of the ontological divide between humans and non-human animals—all of this seemed to remove any remaining "metaphysical obstacles" to a fully materialist explanation of the workings of the natural world, including the human mind. And thus (for many) it removed the remaining obstacles to the wholesale commodification of nature in the name of progress, civilization, and whatever else. The disenchantment of nature was, in this sense, met by a countervailing, secular re-enchantment of the human via the ever-receding liberal horizon: a promise of the future good brought about by the holy trinity of social, technological, and economic progress. The gradual mastery of the energies of the natural world through the auspices of techno-industrial development would one day turn the human into a sorcerer, or even a deity.[42] As Christopher Herbert says of the work of Karl Pearson, "The human being is scientifically reconceived in [Pearson's] *The Grammar of Science* as nothing less than the creator and ruler of the universe—a ductile universe that varies with the passage of human generations—and the author of its laws. Science in Pearson's hyperbolic conception, we may say, becomes the vehicle of 'the deification of man.'"[43] Even Darwin himself extolled humanity's "god-like intellect," its "exalted powers."[44]

But if, as Bilgrami says, the disenchantment of nature seemed to augur a new era of human "term-making," what Darwin's ecological insights also make strikingly clear is that humans do *not* set the terms. What is clear (or should be) is that we are enmeshed in, vulnerable to, at the mercy of, defined by all of the natural forces we would seek to control. Darwin insistently argues that so unfathomably intricate is the character of the natural world—relationships between and among organisms, habitats, and conditions, he writes, radiate outward in "ever-increasing circles of complexity"—that our ape brains, advanced as they are, can never grasp the total picture and are able to comprehend the natural world only in limited, partial, necessarily embodied, and situated ways.[45] In other words, if Darwin's work seems a culminating point in the disenchantment of the natural world, it also turned out to be an inflection point. The very scientific rationality that seemed to promise a future of human "term-making" has also articulated its own insufficiency and the profound irrationality of such an idea. The calls are coming from inside the house. Darwin is key to the story of our culture of denialism, then, not so much because he has been the object of so much reactionary "skepticism" but because his work opens up a profound, unresolved, and unresolvable fault line in the secular, liberal imaginary at the very moment of its ascendancy.

ARTFUL DODGING

Disavowal, John Steiner writes, following Freud, is "artful"—it takes some formal management, some amount of mental artifice, to knowingly not know, to hide things from oneself.[46] It is, therefore, closely aligned with that complex affective-cognitive operation demanded by the realist novel and other forms of bourgeois art: the willing suspension of disbelief. I would note here, in passing, that Mannoni's illustrations for the workings of disavowal are stage magic and theater, both of which often involve an audience that agrees to participate in its own deception.⁴⁷ The nineteenth-century novel, therefore, is ground zero for any consideration of the soft denial of evolution, not just because it arises alongside evolutionary theory, developing out of the same cultural matrix that also gave us Darwinian biology, but also because it commonly asks its readers to practice a form of disavowal in which they knowingly surrender to make-believe for the sake of aesthetic experience. This is "enchantment" in the Nabokovian sense, where we allow the magician-author to show us "the castle of cards become a castle of beautiful steel and glass," even if "we must keep a little aloof" at the same time.⁴⁸ In a less grandiose register, Virginia Woolf describes a similar process by which realist novelists conjure a sense of almost tactile substantiality through the reader's willingness to consent to the illusion: "Each of them assures us that things are precisely as they say they are. What they describe happens actually before our eyes. We get from their novels the same sort of refreshment and delight that we get from seeing something actually happen in the street below."⁴⁹ What I would note here is the emphasis in both cases on the way that (mostly) nineteenth-century realism (Nabokov's examples are Austen and Dickens; Woolf's, Trollope and Maupassant, among others), creates a felt, if factitious, sense of solid ground. For Woolf, the "emphasis is laid upon the very facts that most reassure us of stability in real life, upon money, furniture, food, until we seem wedged among solid objects in a solid universe . . . so that in whatever direction we reach out for assurance we receive it."⁵⁰ The draw is the soothing power of mere security; nothing so grandiose as "enchantment" but an everyday magic that trades upon the same processes: "to believe seems the greatest of all pleasures."⁵¹ Meanwhile, we might note that, however fantastical Nabokov's "steel and glass" castle may be, it inevitably recalls those very real nineteenth-century tributes to industry and commerce: the Crystal Palace and Walter Benjamin's arcades. It suggests, in other words, what Timothy Clark calls "denial in concrete,"⁵² the fantasy of cultural and civilizational perdurability, which we allow ourselves to entertain, even if we also can never quite forget the fact that the whole thing is a house of cards.

This kind of reality effect is precisely what Amitav Ghosh holds up for criticism in *The Great Derangement*. For Ghosh, narrative realism as it has been developed since Jane Austen doesn't just lack a vocabulary for giving expression to environmental crises like climate change; its tendency to abstract the cultural sphere from the natural, to grant outsized significance to "individual moral adventures," over the collective experience of "men in the aggregate," impedes our ability even to conceive of such catastrophic possibilities.[53] Ghosh takes aim at Lyellian uniformitarian geology for the way in which it privileges gradualist conceptions of change and consigns to the realm of the "improbable" more disruptive or discontinuous events.[54] Reginald Terry, writing on that arch-realist Anthony Trollope, makes much the same point as Ghosh, just with a positive spin on the sense of secure reproducibility of the everyday that his novels conjure: "It is an art that depends on time, space, and cumulative effect. . . . The Trollopian mode of undramatic disclosure is a process of gradual revelation through the many insignificant actions and host of tiny observations which, with the author's genial presence as commentator and host, creates a sense of well-being and ease, like opening one's own front door."[55] Although Lyellian geology was a foundational part of the conceptual infrastructure of Darwinian evolution, Darwin himself was also very much attuned to the "uniformitarian expectations" that falsely habituate us to feel that tomorrow will be just like today.[56] As he writes in his *Journal of Researches*, "A bad earthquake at once destroys the oldest associations: the earth, the very emblem of solidity, has moved beneath our feet like a thin crust over a fluid;—one second of time has created in the mind a strange idea of insecurity, which hours of reflection would not have produced."[57] All that is solid moves underfoot, and Darwin notices here—as he frequently does—the sobering discrepancy between the network of associations that nestle us among "solid objects in a solid universe," as Woolf would have it, and the gargantuan forces of the natural world that can level such constructions, material and imaginative, at a moment's notice.[58] An earthquake, in a sense, is an event in which the world of hyperobjects makes itself more immediately available to the senses and the imagination, as the tectonic energies of geological change become palpable through their violent intersection with the supposed stability of the ordinary world. No wonder, then, that when he went to a hydrotherapy resort in 1858, looking to escape the stress of work and temporarily forget the unsettling implications of his own theory, Darwin spent his time getting lost in the latest novels by Trollope, George Eliot, and a host of other now-forgotten three-deckers: "He found them relaxing," his biographer Janet

Browne writes, because "his critical faculties were suspended."[59] For Ghosh, unpredictable events like earthquakes get pushed to the margins of the cultural imaginary by fantasies of solidity and gradualist predictability; soft denial, in short, is embedded in the very conventions and strategic exclusions of narrative realism: "The very gestures with which [the novel] conjures up reality are actually a concealment of the real."[60] Or, as Trollope puts it in his book on Thackeray, "In very truth the realistic novel must not be true,—but just so far removed from truth as to suit the erroneous idea of truth which the reader may be supposed to entertain."[61] Of course, as Ghosh makes clear, the reason readers may be supposed to entertain erroneous ideas about the natural world is that the novel has taught them to do so.

Soft denial is also at work, I would add, in the way those conventions ask the reader to become practiced in the art of self-deception, knowing and not knowing what is real at one and the same time. One could argue that the continual exercise of entering and exiting imagined worlds that require only a strategic half-belief fortifies the tendency to treat stories about the "real" world as yet another imagining, something to be invested in only partially or temporarily. It's worth pointing out in this context that at the very moment it was producing its reality effects, the Victorian novel was *also* busy theorizing its own procedures, self-consciously remarking on the artificiality of the boundaries it set for itself, even while it expected the reader to take such boundaries seriously. Trollope is the high priest of this metafictional necromancy, interrupting his own stories to point to, wink at, and even complain about their adherence to manifestly artificial conventions. As Jacob Romanow demonstrates, such metaleptic interruptions are not at odds with the novel's conjuration of the "real" but components of it: "Metafiction is too often dismissed as a violation of realist epistemology, when it is better understood as a convention of realist generic form."[62] For Romanow, our sense of "real" life is itself an acknowledged fiction, which means that fictions that acknowledge the fictitiousness of the real make themselves more, rather than less, realistic.[63] This helps explain why Trollope can at once be the most realistic of Victorian novelists while also being the one who most frequently and noisily calls attention to the artifice of realism; it is not a contradiction because the two qualities are actually mutually constitutive. To put this in the terms I've been using so far, adding layers of knowingness does not disturb the operation of disavowal but reproduces and thus *fortifies* it. So it is not just that the realist novel depends on its readers to suspend their disbelief in order to enter the fictional world, sympathize with the characters, and so on; it also normalizes that act of suspension by reflecting it back as one of the procedures by which everyday life is constructed by narrators, fictional characters, and "real" people alike. This phenomenon, as we'll see in Chapter 3, is not just true of high-realist works like Trollope's but also of novels that feature excursions into irrealist or fantastical idioms and then strategically, self-consciously conclude by embracing disavowal as a means of reconstructing "the realistic" before our eyes, as it were. *The Time Machine*, *The Island of Dr. Moreau*, and *Heart of Darkness*, even *Wuthering Heights* and *Alice in Wonderland*, operate according to such renormalizing logic, where the ending of the novel allows us to watch a reverse Humpty Dumpty effect: the shattered familiar being put back together again.

In *Human Forms*, Ian Duncan describes the way the nineteenth-century novel responded to—as it also helped make imaginable—a secularized, "evolutionary" conception of the human that was articulated in an array of fields (biology, philosophy, history) and that found its culminating expression in Darwin's mature work. Duncan's brilliant book—to which this one is deeply indebted—shows the twin poles of this response. On the one hand, the novel's famously capacious, protean, open, slowly unfolding form makes it in some ways an ideal means of representing and experimenting with a new "ecological" definition of the human as unfixed, historically in-process, and (potentially) continuous with other forms of life (and nonlife) on the planet. As he puts it, "The novel reorganizes itself as the literary form of the modern scientific conception of a developmental, that is, mutable rather than fixed human nature."[64] On the other hand, narrative realism as a thoroughly bourgeois art form also famously exerts something of a conservative counterpressure in the way it works to stabilize a picture of the "ordinary" world as given, durable, and meaningful, which necessarily entails defining and stabilizing the boundaries of the human. Discussing George Eliot, Duncan writes: "Realism is a sophisticated holding action, the protective reassertion of a formal anthropomorphism in the light of new kinds of knowledge of an inhuman and indeed posthuman world."[65] Advancing and retreating, "lighting out for the Territory ahead" (as Huck Finn puts it), and circling the wagons—we can see in the literary response to evolutionary theory what Fredric Jameson describes as the "antinomies of realism," the necessarily, definitionally irresolvable contradictions internal to its logic and structure. Jameson terms narrative realism a form of "aporetic thinking,"[66] where the forces at play "are never reconciled, never fold back into one another in some ultimate reconciliation and identity."[67] Holding irresolvable contradictions in place without resolution, in Jameson's view, is what makes the realist novel work. It is also of course the definition of "disavowal."

THE LONG, LONG, LONG NINETEENTH CENTURY

What Eliot, Tennyson, and Brontë all recognize is how forms of disavowal might not just function as useful psychological defenses for the individual subject but might also hold a wider kind of positive political and social value for the culture at large. What Eliot, especially, recognizes is that the structuring antinomies Jameson identifies at the heart of bourgeois realism are also the structuring antinomies of the liberal political imaginary, the contrary energy of which is holding the whole thing in place. Fantasy and reality are locked together: interarticulated and reciprocally defining. If there is nothing anyone can do about, say, the inevitable extinction of the human species, then what is the moral or social value in dwelling on it? Or—to put it more strongly—is fully "admitting" such things not a recipe for chaos? One might be an animal, but isn't it better for everyone—socially, ethically, politically, spiritually—if we all agree to aspire to something higher? Even if we know that this "something higher" is just a fiction but behave *as if* it is real, doesn't that, in some sense, bring that reality into being? In any case, aren't some things better ignored in order to just get on with the business of living?

The climate crisis today pretty clearly demonstrates that the answer to that last question is a resounding *no*; that, indeed, the very desire to just get on with the business of living is *itself* part of the problem, especially when we consider how resource intensive that business has become in much of the so-called developed world. Indeed, our contemporary denialist culture seems to operate according to a viciously circular logic—what Klein calls "a doubling down"—where mounting chaos intensifies fear, which, in turn, only increases the allure of the fantasy, even though that fantasy has been exposed as part of the problem.[68] It is a global, extinction-level version of what Lauren

Berlant calls "cruel optimism," a condition where "something you desire is actually an obstacle to your flourishing. . . . The object that draws your attachment actively impedes the aim that brought you to it initially."[69] Such spiraling logic suggests that the familiar trope that the world is "addicted" to fossil fuels appears to be more than simply a metaphor but also a means by which we can see how the psychological, the cultural, the socio-political, and the extractivist-economic are dynamically and mutually reinforcing across a common trans-scalar patterning. For Eliot, Brontë, and Tennyson, there was nothing to be done about the realities described by evolutionary or geological science, and thus it makes sense to imagine that some amount of denial might serve as a useful psychological resource, a coping strategy. We, on the other hand, know well that climate change is a human-caused problem about which there is still *everything* to be done and that denialist defense mechanisms serve, in the long run, only to increase our vulnerability to the very things they purport to fend off. Eliot uses the metaphor of "entrance" and the binary "inside/outside" to refer to the operations of consciousness; what seems clear now is that the metaphor *itself* is leaky and that what she imagines as mental operations have real and unpredictable aggregated material effects. Mind and matter, inside and outside, tenor and vehicle collapse upon each other, just as the various psychic barriers erected to ward off thoughts of climate disaster eventually get breached by unavoidable physical realities: floodwaters overwhelming streets and houses; toxic air from wildfires pouring in through the edges around door frames and windows; excessive heat disabling power grids and cooling systems, making inside as unbearable as out. One difference between nineteenth-century denialism and our own is that the notion of human extinction with which the Victorians struggled undergoes a dramatic transformation when it changes from abstract and distant inevitability to potentially imminent, self-inflicted disaster.

What the climate crisis *also* clearly demonstrates is that many of the basic assumptions of liberal humanism—the preeminence of human concerns over those of non-human animals and environments; the divide between the realms of culture and nature or between human and natural history; the valorization of the individual; the belief in the ultimate efficacy of instrumental rationality, almost always tacitly underwritten by narratives of progress—have played no small part in both bringing various ecological crises about and concealing them from view. The forms of strategic ignorance or willful blindness that seemed to many Victorians as necessary and even harmless defenses against the feared moral and social chaos of a Darwinian cosmos now appear, in the longer view, as a means of both avoiding and abetting far more destructive forms of chaos that have been unleashed by the "humanization" of the planet. Of course, it seems maybe too much like hindsight to blame the Victorians for not seeing this coming, for failing to realize that the forces they were letting loose would have world-altering, maybe world-destroying consequences. At the same time, as I hope to show in the following pages, the denial and disavowal of Darwin was always, from the outset, interwoven with other forms of willful blindness and forgetting that actively obscured the violent, destructive underside of the ascendant liberal order. It was always there to be seen—and thus there to be ignored—in the polluted skies and filthy waterways of an industrializing England; in the immiseration of the laboring poor in city slums and factories; in the shocking forms of cruelty and mass death inflicted upon non-human animals; in the relentless dehumanization, exploitation, and eradication of non-European peoples in the march of empire and capital across the globe. All of these phenomena were objects of denial in various forms, and one of the points of this book is that they must be seen as interarticulated and mutually reinforcing, then and now. The denial of evolution and the denial of climate science are not merely two instances or illustrations of the way discomfiting scientific knowledge gets refused, downplayed, or ignored; they are also two moments in one long denialist story, in which the hard reality of humanity's full embeddedness in the natural world is pitched dramatically against so many long-standing cultural myths and social and economic practices, so many desires and dreams and habits of self-fashioning and exception making, that it remains stubbornly unassimilated—dismissed, evaded, soft-pedaled, conveniently forgotten—even in places where you might expect otherwise.

All of this seems pretty obvious now. But confident as we may be in our superior awareness of everything that is at stake in pushing non-human scales and realities out of mind, it is also unquestionably true that it keeps happening. Most anyone who has spent any time reading about the increasingly dire predictions of climate science has probably, at some point, found ways to throw a veil back over those facts, at least partly or temporarily. I would venture to say that most readers of this book have, at some point, reacted to an especially grim article on climate chaos in much the same the way that Brontë reacts to Martineau's book: I want to know, and I also would rather not. This is even the case for working scientists and ecologists, as the marine biologist Greta Peci puts it: "There is absolutely no doubt at all that it is happening, it is a result of greenhouse gasses and that we need to reduce emissions and adapt asap. On the other hand, I just don't want to believe it either and I really wish it wasn't happening. So I can understand how people deny it."[70] Denialism today is much more extreme in every respect: we are more knowing *and* more avoidant; more clear-sighted and more willfully blind; more acutely aware of environmental crisis and more elaborately girded with various psychological, technological, infrastructural defense mechanisms and means of distraction.

This brings us to the most famous use of the trope of the veil in the nineteenth century, Marx's theorization of commodity fetishism. The commodity, he writes, throws a "mystical veil" over the "life-process of society,"[71] which means, primarily, energy in the form of the laboring power of the working class:

> The capitalist class is constantly giving to the labouring class order-notes, in the form of money, on a portion of the commodities produced by the latter and appropriated by the former. The labourers give these order-notes back just as constantly to the capitalist class, and in this way get their share of their own product. The transaction is veiled by the commodity-form of the product and the money-form of the commodity.[72]

What is veiled is more than just a social relationship; it is also, as Kate Soper puts it, "the productive context" of the commodity in its entirety, including the environmental costs of the manufacturing process.[73] Commodity fetishism creates the cultural fiction of the market as some sort of distinct, quasi-autonomous sphere that operates according to its own inviolable laws; that is, "the economy" itself is fetishized. The naturalization of capitalist abstractions Marx describes produces not only a ready excuse for low wages and the exploitation of workers but also a conception of nature as a mere adjunct or secondary player in the operations of the economic process. The very source of value in a regime of extractive capitalism is, perversely but not surprisingly, what is most thoroughly abstracted and devalued to the point of total disappearance. As we will see in Chapter 5, the eco-theorist Val Plumwood terms this "denied dependency" and describes the way it is bound up with patriarchal culture and the historical "backgrounding" of the labor of women.[74]

In some Marxist criticism, the immense power and cultural saturation of modern advertising and cinema, as well as the vastly increased distance between sites of consumption and production in a "globalized" economy, have created an even more profound structure of ignorance than anything Marx experienced or envisioned in the nineteenth century. Laura Mulvey, invoking Mannoni, argues that "the consumer of commodities is not known to whisper, 'I know very well, but all the same . . .'"[75] That is, unlike the Freudian version of fetishism, where the fetishized object contains some trace or "memorial" of the unsettling knowledge being evaded, the Marxist fetish (she argues) tends to more completely erase any residue of its origin. Soper argues that "the 'veil of ignorance' about the source of the commodity is today that much easier to draw because the prepackaged item on the supermarket shelf is often enough remote by several continents from the misery and pollution of its production, not merely by the distance between high street and local factory."[76] Similarly, Jeremy Seabrook describes the power of modern advertising as the creation of a "nimbus of unknowing":

> One of the most extraordinary by-products of the information-rich societies is the creation of a kind of unknowing, even ignorance, that is strangely at odds with the profuse means of communication that they have at their command. Indeed, some observers have seen in this process a human-made replica of older patterns of natural ignorance, whereby people today have become as unaware of the origin, the violence, the exploitation involved in the production of everyday articles and necessities as the peasantry was once unaware of the forces that governed the rhythm of lives in bondage to the vagaries of the seasons and to the owners of the earth they cultivated. A new and artificial techno-peasantry is in the making: it is to this version of pauperizing people in the rich countries that the advertising industry is dedicated.[77]

Although these are all useful accounts of denial as a materially structured, as well as collectively produced and funded creation of advertising, Hollywood cinema, and global capital, one wonders about the extent to which Mulvey,

Soper, and Seabrook, in attesting to the compelling power of these engines of modern fantasy, overstate the ignorance of consumers and participants; or, to put it differently, fail to register the way in which knowing itself can function as a form of ignorance. Seabrook writes that "the things we can afford are thus sundered from all antecedents, all unpleasant associations," and while this might be true of their actual presentation in commercials or on supermarket shelves, it's also the case that consumers are often *intensely* aware of the social and environmental costs of commodity production.[78] Bruce Robbins argues that "the century in which humanitarianism emerged also bristled with consciousness of how commodities newly enjoyed at home depended on coercion, violence, or mere unpleasantness elsewhere" and notes how keenly aware many consumers are of the horrific working conditions behind the production of, for example, "our beautiful iPhones and iPads."[79] As Slavoj Žižek has it, "The formula of cynicism is no longer the classic Marxian 'they do not know it, but they are doing it'; it is 'they know very well what they are doing, yet they are doing it.'"[80] The entire system of fetishism (of commodities, money, the economy itself), Žižek writes, depends upon the logic of disavowal: "A bourgeois subject knows very well that there is nothing magic about money, that money is just an object which stands for a set of social relations, but he nevertheless acts in real life as if he believed that money is a magic thing."[81] The philosopher Peter Sloterdijk calls this state of mind "enlightened false consciousness," where those who enjoy the material comforts of modernity are intensely and unhappily aware of the manifold horrors of the system from which those comforts arise.[82] For Sloterdijk, enlightened false consciousness is a by-product of the failure of various radical, utopian, emancipatory political projects to achieve any meaningful results in the real world, producing a cynical split between knowledge and everyday life: "It has learned its lessons in enlightenment, but it has not, and probably was not able to, put them into practice. Well-off and miserable at the same time, this consciousness no longer feels affected by any critique of ideology; its falseness is already reflexively buffered."[83] In this sense, commodity production and climate change are not just related insofar as the fossil-fuel intensive nature of the former means it directly contributes to the latter but also in the way that increasing knowledge about the process of production does depressingly little to change the situation. The problem, as Sloterdijk and Žižek define it, is not ignorance of difficult facts but what is to be done about them.

The question of denial thus sets up a complicated relay between the nineteenth-century past and our own present moment. We know so much more than the Victorians did about where this denialist story goes, but that doesn't seem to matter as much as it should. Indeed, one of the problems seems to be with the very idea of *knowing* itself and the tacit cognitivist bias that often accompanies the concept, which imagines that correct action will flow from correct knowledge. Decades of foot-dragging in response to the climate crisis clearly demonstrates that this is very far from the case; that knowing is always defined, constituted, and limited by desire and aversion, the blandishments and threats of the capitalist system, and therefore it can just as easily produce paralysis as action. Maybe more easily. To follow the perverse Möbius strip around one more bend: this means that *knowing*—in the feeling of superiority it can afford; in the illusion of distance it can create; in the sense of critical awareness it can offer—functions as its own form of denial. The paradoxes mount. The political theorist Ben Glasson, for example, suggests there is a difference between *thinking* we know and really knowing and that not knowing which is which becomes an obstacle to effective action:

> We know very well that the earth is under threat, but we do not act on it because we disavow our knowledge. It is what we think we know—but don't really know it at all. Thus we can account for the gap between what subjects say about climate change—that it is an important issue that should be addressed—and what they are prepared to *stand for*.[84]

As Glasson writes, we are "aware but unmoved" when it comes to the climate crisis, by which he means both a refusal to *be* moved (to actually *feel* the problem) and to take any material steps in response to it. The all-important ellipsis at the end of Mannoni's famous phrase "I know well, but all the same . . ." lends it an ironic, even cheeky, knowingness that also suggests an always inconclusive continuation; likewise, its mirrored rhetorical structure—two roughly equal halves poised on either side of the comma—conveys the sense of a see-sawing, forever-unresolved back-and-forth. Motion without movement. The felt sense of both distance and helplessness this produces often expresses itself as a kind of joke, the basic outline of which echoes Mannoni's formulation and which, I would wager, will be familiar to most readers: "I know very well the meat industry is a nightmare; but (all the same) bacon is delicious . . ." Or, "I know Amazon is evil, but it's nice to get packages . . ." Ironic self-consciousness creates a strange kind of permission structure: I hereby acknowledge my complicity, but by acknowledging it, by even recognizing the shameful inadequacy of my response, I allow that complicity to continue. The joke (such as it is) inheres in the stark perversity of the mismatch between the catastrophic (moral, social, environmental) effects of the system and the triumphantly trivial forms of pleasure and gratification that the system provides. What matters so much eclipsed by what matters not at all.

So, disavowal, paradoxically, is an admission of denial. Which means, even more paradoxically, it is also a denial of denial. Because, after all, if I'm admitting I'm denying something, how can I be denying it? It is self-inculpating and self-exculpating at once. Hard denial can be absurd, outrageous, and terrifying in the way it can so successfully sow doubt in established facts. But to identify the creationists, the fundamentalists, the flat-earthers, the so-called climate skeptics as the *real* problem people can be to implicitly credit one's own position as rational, unblinkered, even brave. Such a binarism obscures, as it enables, those softer forms of denial that might be less egregious but are therefore also harder to know what to do with. Disavowal allows one to stand above the conspicuous irrationality of reactionary denialism while still availing oneself of similarly avoidant psychological protections. The very "knowingness" of disavowal functions as a way not to know. It's a meta-update to La Rochefoucauld: the open admission of hypocrisy is the tribute vice pays to virtue.

So whatever sense of critical distance we seem to have achieved in relation to the nineteenth-century past must be tempered with a recognition of all the ways that we continue to dwell in the conceptual, institutional, and material culture of denial that the Victorians helped create, and thus we are still thinking *through* and *with* and *by means of* them whenever we are thinking *about* them. We might take, as Zachary Samalin argues, the edifices and infrastructure erected during the nineteenth century's vast reorganization of physical space—the subway tunnels and sewer systems, hospitals and schools, museums and city streets that are still utilized, visited, traveled on, worked in—as "emblematic of this complex spatio-temporal perpetuation, according to which we continue to inhabit a lifeworld constructed according to a previous era's vision of collective life on the mass scale. Such edifices and the institutions they embody reflect the global reach of nineteenth-century imperialism and a voraciously expanding capitalist economy and culture, within the confines of which a great deal of the world, from San Francisco and São Paulo to London and Kolkata, continues to conduct its daily business."[85] Such edifices speak to the way the built world sets profound, countervailing pressures against any attempts to "realize" our condition of profound environmental crisis, since so much of it functions as an implicit advertisement for its own endless reproducibility, concealing the precarity of its own foundations through a silent, cement-and-steel performance of durability and normalcy. As Timothy Clark puts it,

> "Denial" is less the assumed property of a personality than of the encompassing condition in which it finds itself. Most modern infrastructure in the developed world is, so to speak, denial in concrete, for the distribution of buildings, work places, shopping areas and roads encourages or even enforces certain ways of life, such as private vehicle use, and makes (only temporary) sense in a period of cheap fossil fuel use. For millions of people, the objects and routines of normal, daily life are forms of denial in this extended, only distantly psychological sense, a subtle mix of knowledge, inertia, self-deception, evasion, and material entrapment.[86]

We may know a fossil-fuel-intensive mode of economic organization is enormously destructive, but everything about so many of our daily interactions that take place in, and through, and thanks to, that organization tells us that it "makes . . . sense." The "only temporary" qualifier in Clark's formulation is precisely what gets elided, what makes up the "soft" part of the denial: we *know* it is only temporary, but we are everywhere encouraged to allow ourselves to forget this.

For Mary Poovey, whom Samalin cites in his essay, the roots of what Clark calls "denial in concrete," can be traced back to the utilitarian logic of nineteenth-century industrial capitalism, which worked to "naturalize" abstract ways of imagining and organizing physical environments along with the bodies in them. This served to mystify the "'framework of power' inherent in—and enforced by—the new organization of space."[87] The irony, of course, which seems even more evident now, is that such "naturalization" involved a profound, ecologically myopic *denaturalization*—of bodies, of physical spaces, of ecosystemic pressures and realities—in the service of a disciplinary regime governed by the demands of capital and the economic rationalization of all aspects of life. Abstract space is empty space, and that fantasy of emptiness makes possible (is fueled by) other kinds of fantasies—of order, regularity, predictability, harmony, interchangeability, efficiency—all of which are needed to optimize returns on investment and economic growth. Darwin's work was, to be sure, almost immediately conscripted into the ideological service of capital, but his theorization of ecology refuses every one of the preceding terms, conceiving of physical space as complexly enmeshed and (famously) "entangled." Not empty, but replete. Poovey's interest is not only in how this regime consolidated itself in the nineteenth century but also in how the imperfections in the translation from abstraction to social practice meant it never achieved the "totalized field of power" toward which its inner logic was always urging.[88] The climate crisis presents another way to understand that failure, one that makes starkly apparent how resistant the material world is to such processes of abstraction, how powerless humans are in the face of that resistance, and yet how powerful, and powerfully destructive, the endlessly renewed desire to try to make it submit anyway.

Samalin cites another critic of mass culture, Walter Benjamin, who describes the persistence of the nineteenth century into the present not through the utilitarian logic of economic rationalization but through its flip side: the dreamworlds of consumer capitalism. He writes: "The nineteenth century a spacetime <*Zeitraum*> (a dreamtime <*Zeit-traum*>) in which the individual consciousness more and more secures itself in reflecting, while the collective consciousness sinks into ever deeper sleep."[89] Benjamin identifies here something of the dynamic of disavowal that partly characterizes the response to the climate emergency: an acute, at times even morbid awareness of the crisis on the part of individual subjects (an awareness that "secures" a sense of subjecthood even as it focuses on its own insecurity) matched with the fact that the world collectively is still, as UN Secretary-General António Guterres put it, "sleepwalking to climate catastrophe." For Benjamin, famously, that sleepwalking is the spell cast by the phantasmagorical power of the commodity, which so fully permeates the most intimate dimensions of human life and so entirely dominates experience, that the very idea of critical distance itself is simply another expression of the fantasy: "[Criticism's] day is long past. Criticism is a matter of correct distancing. It was at home in a world where perspectives and prospects counted and where it was still possible to take a standpoint. Now things press too closely on human society. The 'unclouded,' 'innocent' eye has become a lie."[90] Of course, as we've seen, one of the more insidious qualities of soft denial is that one can know something is a lie and believe it anyway; one can acknowledge one's lack of innocence, cop to "guilty" pleasures, understand on some level that mass culture casts a phantasmagoric spell and yet to allow oneself, strategically, to fall (temporarily, ironically, knowingly) under its charismatic sway.

In her celebrated essay "What Is Freedom?" Hannah Arendt describes the way that concept has been gradually stripped of its social and political significance and turned into a question of an individual state of mind. Western philosophy, she writes, "has distorted, instead of clarifying, the very idea of freedom such as it is given in human experience by transposing it from its original field, the realm of politics and human affairs in general, to an inward domain, the will, where it would be open to self-inspection."[91] This "inner freedom," Arendt writes, is a "space into which men may escape from external coercion and *feel* free."[92] For Arendt, the endlessly elaborated internal choreographies of will and self-knowledge are reactions to a coercive external world where the exercise of actual freedom is curtailed or denied and from which, therefore, one feels deeply alienated. Disavowal is one version of this reaction, insofar as it allows the construction of an inner, multilevel space of "escape" where the knowing subject can experience a precarious, illusory sense of control over her engagement with reality—what can be known, when, on what terms, to what extent, and how deeply. Though Arendt has in mind forms of political and social oppression, we can easily see how the feeling of "estrangement from the world," as she calls it, might apply to a sense of entrapment by way of the implacable and inescapable forces of material nature.[93] Alienation *from* the material, in other words, is ironically and dialectically produced by the difficult knowledge of an even more complete enmeshment *within* it. Today, that sense of entrapment presses upon the attention through the increasingly obvious breakdown of earth systems; in the nineteenth century, it arose through the conceptual biologization of all aspects of human experience—from the macrolevel of the historical and social to the microlevel of the affective and behavioral. The nineteenth century closed down some routes of "escape" from the merely material while finding ways to pretend others were still open for business—through the demystifying power of scientific abstraction that was fetishized as a special sign of distinction for select human groups; through fantasies of civilizational development that offered to untangle the human from its now constitutive, definitional entanglements with the non-human; above all, through capitalism's fabrications and false promises, its aggressive commodifying and extractivist drives that offer a waking dream of control over the chaotic materiality Darwin described. Such dreams of control light up and leverage our defense mechanisms while supercharging the threats that make them feel more necessary. The signs pointing to the exit only lead deeper inside.

"We tell ourselves stories in order to live," Joan Didion famously put it.[94] What follows in the next few chapters is a consideration of the stories Darwin and his fellow Victorians told themselves in order to live with unsettling truths about their place in the natural order; how we are, in many ways, still telling ourselves those stories; and what kinds of different stories might now need to be told in order to live.

TWO

THE DENIAL OF DARWIN

ABOUT A YEAR BEFORE *The Origin of Species* was published, Charles Darwin sent a letter to his wife, Emma, from Moor Park, a hydrotherapy retreat in Surrey, where he was staying to recover his health. At times, he felt, that recovery required the willed forgetting of his own work, both the task itself and its implications for his experience of the world around him. Here is what he wrote:

> I strolled a little beyond the glade for an hour & half & enjoyed myself—the fresh yet dark green of the grand Scotch Firs, the brown of the catkins of the old Birches with their white stems & a fringe of distant green from the larches, made an excessively pretty view.—At last I fell fast asleep on the grass & awoke with a chorus of birds singing around me, & squirrels running up the trees & some Woodpeckers laughing, & it was as pleasant a rural scene as I ever saw, & I did not care one penny how any of the beasts or birds had been formed.[1]

As the critic Adam Phillips writes about this passage, "Nature can be infinitely reassuring, a virtual idyll, if we don't think about its history. The delight, the pastoral, is sustained through the denial of origins. . . . We are always relaxing in the killing-fields."[2] Darwin was intensely aware of the "virtual" quality of this idyll and how a previously uncomplicated, heart-leaping-up response to the landscape is now achievable for him only via a state of temporary forgetting. For a moment, he does not care about the organizing concern of his entire professional life, does not know what he knows. It's a brief aside, but it's in keeping with the rest of Darwin's time at Moor Park. On the very doorstep of the publication of his magnum opus, he sought relief from his debilitating nausea, the pressures of writing, and the very contents of his own thoughts. Stomach and mind, thoughts and nerves, physical and mental stress—for Darwin these are all intensely interconnected and bound up with the implications of his theory both for his professional life and for his own experience of being alive. "My God," he wrote to J. D. Hooker, about six months before *The Origin of Species* appeared, "how I long for my stomachs' sake to wash my hands of it,—for at least one long spell."[3] The desire for escape from the pains and limits of his own animal body was inseparable from the ideas he was developing—that we are all, the entire species, entirely and inescapably animal.

So, Moor Park was an escape, but there is no escape. In many ways, the *whole point* of his theory is exactly that—there is no organism exempt from the pressures of selection and survival; there is no place on earth not organized by those pressures; there is no experience, no form of beauty, no phenomenon that cannot be explained by reference to what Janet Browne calls "the black forces beneath nature's surface."[4] As Phillips puts it, Darwin's theory meant "there were no pastoral retreats now, no plausible refuges from the struggle: war is our continuity."[5] There is something more than a little poignant about a man seeking out a pastoral retreat to escape his own discovery that there is no such thing.

In a way, then, denial is at the origin of *The Origin*. It is not only something the author felt keenly during its composition; it is a structure of feeling he considers, at a number of junctures over the course of his scientific career, as both hindrance and help to the widespread acceptance of his theory. In a section of his *Journal of Researches* (1839) he writes about an array of intellectual and emotional reactions to the idea of ubiquitous struggle in nature and the problem of extinction:

> Certainly, no fact in the long history of the world is so startling as the wide and repeated exterminations of its inhabitants. Nevertheless, if we consider the subject under another point of view, it will appear less perplexing. We do

not steadily bear in mind, how profoundly ignorant we are of the conditions of existence of every animal; nor do we always remember, that some check is constantly preventing the too rapid increase of every organized being left in a state of nature. The supply of food, on an average, remains constant; yet the tendency in every animal to increase by propagation is geometrical; and its surprising effects have nowhere been more astonishingly shown, than in the case of the European animals run wild during the last few centuries in America. Every animal in a state of nature regularly breeds; yet in a species long established, any *great* increase in numbers is obviously impossible, and must be checked by some means.[6]

Darwin is standing on the doorstep of natural selection, ushered there by Malthus and his principle of population. But in the various contraries of thought and feeling described, we can also see how these ideas are attended by the operations of forgetting, inattention, and other denial-adjacent states of mind. Extinction, Darwin says, is the most startling fact *in the history of the world*. And yet, it is also not all that surprising, as long as we keep in mind something we already know: that animals are dying in cataclysmic numbers all the time. Any population, even a thriving one, is continuously emerging out of a measureless, unseen circle of mass failure and death: "an action going on, on every side of us, and yet barely appreciable," as he calls the "checks" to population.[7] So while there is nothing surprising at all, he writes, when we encounter a "rare" species in nature, that very rarity means that species is an environmental change or two away from disappearing altogether: "Why should we feel such great astonishment at the rarity being carried a step further to extinction?"[8] The struggle is constant, the purchase tenuous, especially when imagined on the proper scale. Extinction is not some surprising, anomalous event but an ever-present possibility. Yet, because this fundamental reality of life is "barely appreciable," not directly available to the senses, it takes a willing act of imagination, the conscious adoption of "another point of view," not to forget it. Without that act, without the effort it requires to "steadily" keep it in mind, its effects can become strange and startling, unnecessarily mysterious, and therefore productive of all manner of misunderstanding and error.

Twenty-five years later, the problem of avoidance, forgetting, and denial are still very much on Darwin's mind as he's writing *The Origin of Species*. In the third chapter, "The Struggle for Existence," he reworks the previous passage from his *Journal* to issue a reminder about bearing in mind the known-but-forgotten realities of nature:

> We behold the face of nature bright with gladness, we often see superabundance of food; we do not see, or we forget, that the birds which are idly singing round us mostly live on insects or seeds, and are thus constantly destroying life; or we forget how largely these songsters, or their eggs, or their nestlings, are destroyed by birds and beasts of prey; we do not always bear in mind, that though food may be now superabundant, it is not so at all seasons of each recurring year.[9]

For Darwin's purposes, this cultural habit of "forgetting" things we already know represents both impediment and opportunity for his theory. On the one hand, there is something obviously aversive about destruction, famine, and extinction, and he will need to overcome that aversion to get his audience on board with his arguments. On the other hand, the work of overcoming it involves not convincing them of something entirely new but reminding them of what they already know. Getting them to "admit" what is already inside. The question of struggle and extinction is foundational to the entire project, so succeeding in that work of overcoming and remembering would represent a giant step toward bringing the entire evolutionary picture into focus.

The anthropomorphic idyll Darwin describes at Moor Park, complete with birds pouring forth in a "chorus" and woodpeckers "laughing," is here organized into the humanized "face" of nature. The metaphor is familiar and conventional and also, for that reason, a conspicuous oversimplification. This becomes even clearer a few pages later, when he returns to it in a startling new way: "The face of Nature may be compared to a yielding surface, with ten thousand sharp wedges packed close together and driven inwards by incessant blows, sometimes one wedge being struck, and then another with greater force."[10] The vision of ten thousand wedges being hammered into a surface isn't just an anti-aesthetic gesture; it verges on anti-representational: "hard to picture" on more levels than one. Being willing to "face" nature involves defacing it, which, paradoxically, allows something like its "truer," more abstract, more inhuman face to emerge. Throughout the book, Darwin pursues an array of strategies to encourage his readers to remember, to bear in mind, to see past the pleasing surface, to face what they need to face. The attack of the wedges is Darwin at his most confrontational and estranging, not simply pointing out an obfuscating metaphor or misleading linguistic convention but aggressively destroying it. We might note that he excised this strange passage in later editions, recalibrating, perhaps, the extent to which he was willing to get up in his readers' faces.

Near the end of the book, the approach to the question of denial shifts once more, and in a crucial way. When Darwin returns to the idea of forgetting or failing to see, the target is no longer the widespread cultural habit of denialism of which he himself is sometimes guilty but a much more specific group of deniers: the old guard of scientific naturalists whose hostility to his work he identifies as motivated ignorance: "Nature may be said to have taken pains to reveal, by rudimentary organs and by homologous structures, her scheme of modification, which it seems that we wilfully will not understand."[11] Such was the strength of the case for evolution, so overwhelming was the evidence for the vast age of the earth, the mutability of species, and other foundational components of the paradigm that certain forms of disagreement must be understood not as expressions of legitimate dissent but as defensive refusal, deliberate obfuscation. He writes: "It is so easy to hide our ignorance under such expressions as the 'plan of creation,' 'unity of design,' &c., and to think that we give an explanation when we only restate a fact."[12] In the passage on extinction from the *Journal of Researches* and the one on birdsong from *The Origin of Species*, the collectively forgetting "we" is companionable and inclusive; here it is accusatory and line drawing. *We* becomes *they*: the deniers, busy begging questions and passing off tautologies as arguments, both because they lack "flexibility of mind" and because they seek the path of least resistance.[13] In the early chapters, it was still okay to talk about "forgetting" and admit one's difficulty keeping certain things in mind; now that we've reached the end and the case has been made, the time for denial is over. The plot of *The Origin* is one of emerging out of a state of half-seen, half-avoided truths into a state where that which must be faced can be. Darwin looks forward, he says, "to the future, to young and rising naturalists, who will be able to view both sides of the question with impartiality. Whoever is led to believe that species are mutable will do good service by conscientiously expressing his conviction; for only thus can the load of prejudice by which this subject is overwhelmed be removed."[14] The lines are drawn, the sides are clear: in the future these things will be judged with "impartiality," free of emotional obstacles and the desire to forget.

In one sense Darwin is of course entirely correct that he would turn out to be mostly right and that future generations of scientists would take up his ideas and run with them. But with Dillard, Gould, and Carson's voices in our ears, this description of a future in which evolution could be discussed in an impartial, prejudice-free way sounds way too optimistic. Indeed, we might note that, a few pages later, Darwin himself provides something of a template for how his ideas would get "slanted" and spun—to borrow Stephen Jay Gould's terminology—into something more pleasing: a positive, upward-tending vision of evolution.

> It is interesting to contemplate an entangled bank, clothed with many plants of many kinds, with birds singing on the bushes, with various insects flitting about, and with worms crawling through the damp earth, and to reflect that these elaborately constructed forms, so different from each other, and dependent on each other in so complex a manner, have all been produced by laws acting around us. These laws, taken in the largest sense, being Growth with Reproduction; Inheritance which is almost implied by reproduction; Variability from the indirect and direct action of the external conditions of life, and from use and disuse; a Ratio of Increase so high as to lead to a Struggle for Life, and as a consequence to Natural Selection, entailing Divergence of Character and the Extinction of less-improved forms. Thus, from the war of nature, from famine and death, the most exalted object which we are capable of conceiving, namely the production of the higher animals, directly follows. There is grandeur in this view of life, with its several powers, having been originally breathed into a few forms or into one; and that, whilst this planet has gone cycling on according to the fixed law of gravity, from so simple a beginning endless forms most beautiful and most wonderful have been, and are being, evolved.[15]

Having asked us to bear in mind the aversive things we would rather forget—constant struggle, mass death, extinction—Darwin here rewards us with the story of a brighter version of the natural world ultimately emerging out of all the darkness. We travel from mere surface observation, a pleasing idyll of birdsong, down into the underground lives of insects and worms, and then past that into a supersensible, abstract world of forms, complexities, and laws. The mere visual confusion of the "entangled" scene now gets revealed as a system of intricate, multifactor, and reciprocally defining relationships, but this, in turn, uncovers some of that system's horrifying realities: the capitalized Struggle and Extinction, along with war, famine, and death. Fortunately, however, this journey into the underworld is not our final destination; instead, as Phillips writes, we end with a "consoling vindication of the pastoral of hierarchy and the founding Father. If the question of what nature is like has now been answered—it is like a war, in which there is famine and death—there is a reassuring progress myth underpinning it."[16] The story offers a kind of recapitulative narrative arc conflating the development of Darwin's own inquiry with the development of life itself. The progress of science morphs into evolution *itself* as progress, as if the scientific project is merely a synecdoche for the material phenomenon that project seeks to decipher. Both unfold through the same upward-tending logic. It's a perfect distillation of what Elizabeth Povinelli calls "disavowal function of the liberal horizon," which, as I discussed in the previous chapter, "naturalizes systemic social harm in the *social tense* of the coming good."[17] In this case, the stochastic violence of nature gradually gathers into meaningfulness through an imagined end point that is both revealed and made possible by the advance of scientific reasoning. Our confrontation with denialism has been rewarded; our worst fears have not come to pass. In this way, we can see an example of how the now outwardly directed, line-drawing attack on denialism of others has provided cover for other kinds of evasions.

When Darwin goes after the "ignorance" of those naturalists arguing for the "plan of creation" or "unity of design," he attacks a metaphysical framework premised upon providential final causes; but instead of jettisoning teleology, he refashions it into what the historian of science David Kohn calls "teleology without purpose" and makes his theory amenable to an ascendant secular, liberal imaginary.[18] Such a phrase bristles with paradox, as do many of the words and phrases Darwin presses into ambiguous service to describe the workings of the evolutionary process. What do terms like "exalted" and "higher" even signify in a natural world in which survival and reproduction are the only imperatives and measures of success? Neanderthal Man is extinct. Can he be said in any meaningful sense to be more "highly" evolved than the roach and slug species that have outlasted him? As Gould shows, Darwin was well aware of the anthropocentric biases in these and other terms used throughout *The Origin of Species*; for a time, he even resisted the word "evolution" itself, preferring "descent with modification" as a more neutral descriptor that did not come laden with suggestions of an upward trajectory.[19] Indeed, in his copy of Robert Chambers's popular and much-derided evolutionary work *Vestiges of the Natural History of Creation* (1844) Darwin jotted the note: "Never use the words higher & lower—use more complicated, as the fish type (& not a mere repetition of parts) where cartilaginous forms are higher for being nearer reptiles & consequently mammalia."[20] Not surprisingly, historians of science argue over whether this comment means Darwin was committed to an antiprogressionist position or whether he was simply reminding himself to take care with loaded words.[21] What seems beyond dispute is that he was well aware of the implications of this rhetoric and used it anyway. "Higher," as it appears at the end of *The Origin*, cannot be said to merely signify "more complicated" when used in tandem with a word like "exalted" and imagined as the consoling outcome of eons of cosmic warfare. Something of the logic of disavowal is thus woven into the very fabric of the text as it rises to its conclusion. Phillips writes: "As Darwin well knew, it was precisely this kind of hierarchy, this way of thinking about nature, that his work was questioning. . . . In actuality Darwin was describing a war in which there is no 'great work' or 'exalted object.'"[22] Indeed, Phillips argues, this disjunction between rhetoric and content characterizes Darwin's work as a whole: "His language hints at a politics and a theology that the content and context of the work seem to disavow."[23] It's not that Darwin's language just candy-coats the bitter pill; it sometimes makes you doubt whether you even need to take your medicine.

If *The Origin of Species* delivered its most profound blow to religious dogmatism, its ideas also struck in quieter and less obvious ways against almost all of the foundational concepts and assumptions of secular humanism: the belief in individual autonomy; sovereign rationality; human exceptionalism; the divide between culture and nature; the progressive overcoming of the limits and pressures of the natural world. Decisively undercutting any basis for the idea of an immortal human soul would appear to have mostly theological ramifications until we consider how many of those tenets of secular humanism also, if less explicitly and less programmatically, depend upon the idea of a qualitative difference between human beings and non-human animals. As Ian Duncan writes about the eighteenth-century French naturalist George LeClerc, Comte de Buffon,

> Buffon erects a defensive barrier around man in two essays which address the problematic situation of the species—at once inside and outside the order of nature. In the first of these essays, Buffon contends that man's physiological constitution places him among the animals while the soul, an inner immortal substance manifest in the work of reason, sets him absolutely apart. In formal terms, in short, man is a monster—a cross between alien orders of being. In the second of these essays, "Homo Duplex," Buffon characterizes the taxonomic division between man and the other animals as an ontological state of "perpetual war" within man himself.[24]

Ernst Mayr remarks that "Buffon was not an evolutionist, and yet . . . he was the father of evolutionism" for raising "extensive discussions of the origin of the earth in general, and of sedimentary rocks in particular . . . the problem of the extinction of species of animals . . . and the problem of problems, the establishment of reproductive isolation (as we would now call it) between incipient species."[25] Yet, this forward-thinking evolutionism needed to be accompanied by a "defensive" position that could maintain the privileged integrity of the human species; such a position, in other words, is deeply ingrained in the history of the science of evolution, not to mention in the work of Descartes, Kant, Locke, Bacon, and any number of other major figures of the Western secular humanist tradition. Thus, as Duncan shows, "Homo duplex" remains a kind of metaphysical touchstone even after the ontological justification for his existence—the radical, qualitative divide between humans and animals—had been dispensed with:

> In *The Descent of Man* Darwin would propose a decisive closure of those divisions and caesurae, a resolution of the chimerical figure of "Homo duplex," with a biological accounting of all aspects of the human in which reason, sentiment, taste, morality and religion are subsumed within a unified natural order. . . . At the same time (*circa* 1870), the newly established science

of anthropology would found its disciplinary autonomy upon the disputed border between nature and culture—in effect saving and reinstating "Homo duplex."[26]

Decisively closed, but also reinstated; chimerical, but foundational. Notice how Duncan's own language bridging these opposing ideas ("at the same time") echoes Mannoni's formulation of disavowal ("all the same . . ."), suggesting a similarly paradoxical knowing-and-not-knowing position at the very root of the nascent discipline of evolutionary anthropology. This tension runs through much of the scientific writing on evolution as well: a kind of practiced soft denialism that is often explicitly pitched against—but is implicitly made possible by—the purveyors of "hard" denialism who were identified as the real enemy.

Consider T. H. Huxley's famous line in his review of *The Origin of Species* in G. H. Lewes's *Westminster Review* in 1870, which praises Darwin's book as "a veritable Whitworth gun in the armoury of liberalism."[27] The very vividness of such a metaphor throws into relief the same dynamics that are much more subtly rendered in Darwin's concluding chapter. That is, insofar as the "enemy" can be identified as those Darwin describes as mired in "ignorance," clinging to thoroughly discredited notions of providential design, *The Origin of Species* appears to Huxley as the perfect weapon. The measured notes of optimism sounded in Darwin's final paragraphs swell into the sounds of a victory march: "Extinguished theologians lie about the cradle of every science as the strangled snakes beside that of Hercules, and history records that whenever science and orthodoxy have been fairly opposed, the latter has been forced to retire from the lists, bleeding and crushed, if not annihilated; scotched, if not slain."[28] But obscured by such bellicose imagery is the way in which such a weapon is not really in anyone's control and, indeed, can be turned upon its wielder. *The Origin of Species* is better described not as a machine gun but as the proverbial double-edged sword, bloodying up the humanist wielder, just in less obviously apparent ways. This shows itself in subtle fissures and ripples in the language; unstable ironies and invisible air quotes; tensions between explication and implication; or, as Phillips has it, the misalignment of generic codes and scientific concepts as "the text itself begins to aspire to something at odds with its genre."[29] The kinds of cracks visible in *The Origin*—between content and form, between a grim scientific picture and the more reassuring "hints" layered into the rhetoric—widen into open fractures in much of the evolutionary discourse of the period. Huxley's lifelong project, to articulate, in Adrian Desmond's terms, "a rounded scientific humanism," is internally riven by the different directions in which those two key terms were pulling: evolutionary science working to dismantle the kind of species exceptionalism upon which humanism was premised.[30] As George Levine puts it, "To write like a scientific naturalist is to create a world that is fundamentally self-contradictory."[31]

Not surprisingly, these tensions can be felt everywhere in Huxley's writing. Consider his comments in *Evidence as to Man's Place in Nature* that, thanks to the retentive and directive operations of culture, which "accumulate" and "organize" experience, "man" "stands raised . . . as on a mountaintop, far above the level of his humble fellows, and transfigured from his grosser nature by reflecting, here and there, a ray from the infinite source of truth."[32] It's worth mentioning that "reflecting" here does double duty, signifying both the passive property of a surface redirecting light and the active work of a human intelligence that is capable of "reflection"—that is, able to stand back and consider the cosmos and its own place within it. But further, we might ask what a word like "transfigured" even means in this context. On one level, Huxley is deliberately playing with mismatching linguistic codes and frames of reference, co-opting the language and mountaintop imagery of the Transfiguration from the New Testament to poke a finger in the eye of his religious-minded opponents. He isn't saying humans are transfigured in some supernatural sense. And yet, he isn't *not* saying that, since the word is unavoidably loaded with exactly those associations. Huxley is winking at the reader, but he isn't being *merely* parodic. He is, on some level and however vaguely, attempting to suggest some ground of differentiation between humans and non-human animals even as he, at the same time, is vigorously attacking his religious opponents for clinging to such distinctions. "Transfigured," in other words, is simultaneously used and crossed out, offered and withdrawn—disavowed. Defining "man's place in nature"—*the whole point of the book*—leads us to a non-place, a mere trope, a fabrication of language.[33]

We see these kinds of rhetorical gambits whenever the subject of non-human animals arises. Earlier in the same chapter, Huxley writes:

But it is not I who seek to base Man's dignity upon his great toe or insinuate that we are lost if an Ape has a hippocampus minor. On the contrary, I have done my best to sweep away this vanity. I have endeavoured to show that no absolute structural line of demarcation, wider than that between the animals which immediately succeed us in the scale, can be drawn between the animals and ourselves; and I may add the expression of my belief that the attempt to draw a physical distinction is equally futile, and that even the highest faculties of feeling and of intellect begin to germinate in lower forms of life. At the same time, no one is more strongly convinced than I am of the vastness of the gulf between civilized man and the brutes; or is more certain that whether *from* them or not, he is assuredly not *of* them. No one is less disposed to think lightly of the present dignity, or despairingly of the future hopes, of the only consciously intelligent denizen of this world.[34]

The "hippocampus minor" reference is a nod to a debate between Huxley and Richard Owen, a fellow naturalist and vocal opponent of Darwin's theory, who (incorrectly and contrary to available anatomical evidence) claimed that only *Homo sapiens* possessed this specific part of the brain. Owen argued that such a feature proved humans could be differentiated, finally and decisively, from the great apes. Huxley's discussion of this controversy presents us with the familiar dynamic: Owen is the denier ("guilty of wilful & deliberate falsehood," as Huxley puts it elsewhere), who simply refuses to credit the plain and obvious evidence.[35] In contrast, Huxley positions himself as willing to fully accept the situation, to reject the "vanity" and "futility" involved in trying to hold to any hard-and-fast distinctions between humans and non-humans. But if there are no hard distinctions, there are certainly soft ones. Huxley discards any structural lines of demarcation, and yet "at the same time" (all the same . . .) he posits a "vast gulf." Where there once had been a hippocampus minor holding the line, we now have the dubious reassurances of rhetoric—"whether *from* them or not, he is assuredly not *of* them." The differential weight of the prepositions "from" and "of" (the emphasis is in the original) and the subtly disclaiming work of "whether" are now what must hold the human/non-human line. It is, to borrow a phrase from the historian Walter Johnson, "beguilingly otiose."[36] In other words, in the process of decrying Owen's hard denial, Huxley opens up a space for his own softer version. There are no qualitative distinctions; all the same, there kind of are. The binary terms that are no longer clearly distinguishable are nevertheless still treated *as if* they hold. As Kay Anderson puts it,

> For Darwin and perhaps especially Alfred Wallace, that categorical distinction of human and animal with which western philosophy and theology had grown up, implied a structural gap that now had to be bridged. The fossil evidence of human continuity with the apes required it. And yet, . . . what is fascinating about this period is that the full implications of the "fact" of human-animal evolutionary continuity appears not to have been fully drawn out or confronted. The *felt* sense of categorical distinction of the human from the non-human persisted.[37]

Anderson's "and yet" suggests something of the tension I discussed in the previous chapter: the sense of a divide between knowledge and feeling, or, to return to Sedgwick's terms, between what is known to be "real" and what is actually "realized"—accepted, borne steadily in mind, made integral to how one sees oneself and the world.

Maintaining the line between human and non-human animals allowed for the zones of culture and nature to be differentiated into oppositional spaces, despite all the ways in which Darwin's work was otherwise obliterating that very opposition. Where so-called Social Darwinists like Herbert Spencer argued that all human social institutions were themselves merely expressions of natural forces (and thus should remain free from state regulation), Huxley maintained that civilization represented a crucial stronghold against the struggle, violence, and untrammeled competition that characterized the so-called state of nature. As Piers Hale argues, Huxley believed that

> ethics were of entirely human origin. Thus, in his attempts to discredit both Christian and socialist claims to an *a priori* ethics in nature, and at the same time distance himself from Spencer's individualism, Huxley was forced to place ethics *a priori* in the historical development of society, and crucially, as something that separated humanity from the natural world.[38]

Although Spencer's Social Darwinism was more notorious, Huxley's attempt to erect a barrier between non-human and human domains was more in step with the main currents of Victorian social and political thought and had its own unsavory implications. Historian of science Donald Worster writes:

> The unequivocal defense of civilization as rule by the strongest, common though it was in popular philosophy, was never wholly persuasive to the Victorian mind. It linked man and nature, civilization and savagery, too closely on a continuum of sheer power, when what the Victorians most desired was to establish a vast moral distance between these poles. A second strategy, which bypassed this objection, was to defend civilization as the necessary, rational management of nature.[39]

As I discuss more thoroughly later, the ecological and imperial imaginaries are mutually constitutive and necessarily connected here: civilizations are defined by the degree to which they are at the mercy of natural forces, and therefore the ability to manage those forces (both individually and socially) is understood as both expressive and productive of racial superiority. Despite the powerfully deconstructive potential of Darwin's conceptualization of the evolution of *Homo sapiens* from non-human animals, one influential interpretation of this idea was that it mandated that (white, European) humans ought to take greater command of the natural environment to further distinguish themselves from it and from those who lacked the technological means to do so. So, although we might expect that Darwin's discussion of the complex interconnectedness of all life would raise serious concerns about industrial growth and the rippling consequences of things like invasive mining practices, large-scale forest clearing, and increasing fossil-fuel consumption, in fact, it became in many influential quarters an important argument *for that very regime.*

Unsurprisingly, then, Huxley's vision of modern civilization positions it as simultaneously within and without nature, a product of its forces and a zone crucially set apart from them. On some level, he can thread this needle because his work, like Darwin's, features different (though not incompatible) visions of nature that make possible different conceptualizations of human agency. Even more than Darwin, Huxley tends to imagine "Nature" as a set of abstract laws—what he often calls "the cosmic process"—that brings the living world into being and yet remains set apart from it. That remote, abstract thing "the cosmos" becomes the backdrop against which the human must define itself and maintain its material and imaginative boundaries. For Huxley, civilization is to be thought of as a kind of "garden"—this is the controlling metaphor of *Evolution and Ethics*—a place that is natural and yet also artificial, crucially walled off from the "antagonistic" and "hostile" matrix of forces that are always threatening to (and will one day) overwhelm it.[40] In such a space, he says, people can manage and order and exclude the "coarse struggle for existence" by imposing whatever regime of artificial selection best aligns with their moral and political ideals and advances their notion of social well-being.[41] But that system is *itself* unavoidably embedded within the Darwinian paradigm: "But the Eden would have its serpent, and a very subtle beast too. Man shares with the rest of the living world the mighty instinct of reproduction and its consequence, the tendency to multiply with great rapidity. The better the measures of the administrator achieved their object, the more completely the destructive agencies of the state of nature were defeated, the less would that multiplication be checked."[42] By cordoning off a world free of the pressures of natural selection, the managers of the garden only delay its operation. Or, to put it more precisely and paradoxically, the more natural selection is kept outside the garden walls, the more serious a threat it becomes inside. Exclusion is inclusion; success produces failure. The fight may be unwinnable, but it still must be had: "Let us understand, once for all, that the ethical progress of society depends, not on imitating the cosmic process, still less in running away from it, but in combating it."[43] In Huxley's telling, the story of (European) civilization is a tragic drama of sorts, an ennobling struggle against a blind process to which it must one day succumb. One can appreciate the difficult line Huxley has to walk here—trying to define the human sphere in Darwinist but not Social Darwinist terms—while also recognizing that there is something deeply suspect from an environmental point of view about conceptualizing the relationship between nature and culture as a giant cosmic war.

We find everywhere in Huxley's writings assertions of European superiority, descriptions of the practical and moral need for industrial development, and the urge "to subdue nature to . . . higher ends."[44] This paradoxical inside-outside conception of human civilization plays up the sense of divergent "realities" we see in *The Origin of Species:* the usual linguistic conventions and forms of forgetting that conjure the beneficent "face" of nature versus what we know is actually going on under the ground and in the tangle. The tension Darwin suggests between the conventional and the natural becomes in Huxley an explicit opposition between states of art and nature:

> That which lies before the human race is a constant struggle to maintain and improve, in opposition to the State of Nature, the State of Art of an organized polity; in which, and by which, man may develop a worthy civilization, capable of maintaining and constantly improving itself, until the evolution of our globe shall have entered so far upon its downward course that the cosmic process resumes its sway; and, once more, the State of Nature prevails over the surface of our planet.[45]

Such a passage represents both an elaboration of Darwin's understanding of opposing realities and a significant repurposing of it. In 1859, when *The Origin of Species* was first published, Darwin demystified nature in order to make his so-called dangerous idea seem compatible with a widely acknowledged, if "forgotten," reality. When Huxley poses the opposition between Nature and Art, it is 1893: The Darwinian revolution is in full swing, and Spencerian arguments about the "state of nature" in social policy threaten to justify laissez-faire economic individualism as the natural order of things. The task at hand, as Huxley saw it, was no longer to demystify the garden by urging people to remember but to refortify its walls to allow them to temporarily forget. Huxley doesn't want to return to a sentimentalized vision of nature but to make possible a strategic and knowing resubscription to artifice. Willing illusion brings into being the reality it simulates; belief that there is such a thing as progress is what makes the thing we call progress possible, and "civilization" exists only insofar as people think it does. This has obvious implications for the imperial context, and more on that in the next section. But here, it's worth pointing out that, understandable as Huxley's motivations may be, we can see how his dichotomy makes a kind of proleptic apology for the Anthropocene by imagining the total humanization of the earth as a consummation devoutly to be wished. The war here is territorial, a fight for the "surface of the planet," and the forces of destabilization and disintegration are not what humans unleash but what they hold back.

WALLACE'S PLOTS

The foregoing helps makes sense of something that otherwise seems deeply counterintuitive: Despite his robust theorization of infinitely complex forms of interactivity and his own firsthand experience of the human impact upon natural environments, despite his absolutely central place today in the annals of Western ecological thought, Darwin does not seem like much of an "environmentalist" in the way we use that term today. The human is an important, but quiet, and largely benign presence in most of the well-known moments of ecological storytelling in *The Origin of Species*. He describes the aftereffects of clear-cutting forests in America; the unintended consequences of enclosing portions of the heath in Staffordshire; the entirely unplanned and unexpected effects of human settlements and the introduction of house cats on the insect and plant life in a region. And yet, in all these cases, human agency is downplayed rhetorically through passive constructions ("when an American forest is cut down"; "large spaces have been enclosed.")[46] More important, the effects of these passive ecological interventions are often coded as advantageous or at least neutral: the forests grow back; the enclosure causes plant life to flourish; the house cats make the clover bloom. Environmental historian Richard Grove argues that although *The Origin of Species* was important for many nineteenth-century environmental activists, "Darwin was himself remarkably unconcerned, or seemingly so, to advocate the prevention of extinctions or the preservation of forests."[47] And historian of science Nancy Stepan goes so far as to say that "Darwin drew no larger conclusions from this knowledge of interdependency, remaining unconcerned about conservation issues."[48] "Unconcerned" is perhaps strong, but it speaks to a sense we get when returning to Darwin's major writings that he does not seem as attuned as one might expect to the "green" implications of his own work or as alarmed by the damage being visited upon actually existing environments at the hands of capital and empire. Indeed, at times, he dismisses concerns about human impact: "Man has no power of altering the absolute conditions of life; he cannot change the climate of any country; he adds no new element to the soil; but he can remove an animal or plant from one climate or soil to another, and give it food on which it did not subsist in its natural state. It is an error to speak of man 'tampering with nature' and causing variability."[49] Although, to be sure, the terms he uses in this passage arise out of a very different context, it's hard today to read the claim that humans "cannot change the climate of any country" and not register the grim irony in its sense of assurance.

As we've seen with Huxley, many of Darwin's fellow Victorian scientists go far beyond the kind of benign neglect Stepan describes and at times write about the natural world in shockingly oppositional terms, as something to be attacked, subdued, managed, and dominated. As I've been suggesting, at least some of this has to do with the radical challenge to human exceptionalism posed by Darwin's work, a challenge that would seem to demand much greater humility and care in our approach to the natural world, a much keener recognition of the baneful, ramifying, unforeseeable consequences of industrial growth, water pollution, clear-cutting, strip mining, fossil-fuel combustion, and any number of other environment-altering activities. And yet it seemed, in many quarters, to do something like the opposite. By shrinking the human down to a speck in the annals of geological time, evolutionary discourse both minimized its potential impact on the planet and provided the grounds for some to engineer their way back into a sense of significance. If humans had been decentered in some profound ontological sense, the answer was, for many, to reclaim it by other means, through other discursive and material strategies.

But the fact that a powerful ecological critique of Western industrialism was there to be developed from Victorian evolutionary discourse becomes clear when we turn to Darwin's rival and "co-discoverer" of evolution by natural selection, Alfred Russel Wallace. Unlike Darwin, Wallace paid close attention to the kinds of damage being inflicted upon tropical regions by a ruthlessly extractivist imperial system and, over the course of his career, became an outspoken critic of capitalism. In *The Malay Archipelago*, for example, he describes his experience tracking down a "King Bird of Paradise" in the Aru Islands in present-day Indonesia:

> Should civilized man ever reach these distant lands, and bring moral, intellectual, and physical light into the recesses of these virgin forests, we may be sure that he will so disturb the nicely-balanced relations of organic and inorganic nature as to cause the disappearance, and finally the extinction, of these very beings whose wonderful structure and beauty he alone is fitted

to appreciate and enjoy. This consideration must surely tell us that all living things were *not* made for man.[50]

Wallace is struggling here with anthropocentric thinking, as he wants to maintain the significance of human consciousness ("he alone is fitted to appreciate") while insisting that appreciation does not equal ownership. Look but don't touch; or, more precisely, looking doesn't give you the right to touch. Wallace's human-centered aestheticism sits uneasily with the nascent theory of sexual selection, but it carries the force of an ethical or political commitment against the wholesale commodification of the natural world. To be sure, his description of the "moral, intellectual, and physical light" of civilization is not altogether ironic—one senses he still, on some level, believes in such things. But much more than Darwin, Wallace is willing to look squarely at the damage being visited upon non-human life and the integrity of complex ecosystems by the advance of capital and empire across the globe. This critique grows only more trenchant over the course of his writing life. Consider this passage from his archly titled retrospective *The Wonderful Century*, in which he turns his focus to environmental matters closer to home and considers the dynamics of "climate change" in England:

> Some years back one of our gardening periodicals obtained from gardeners of forty or fifty years' experience a body of facts clearly indicating a comparatively recent change of climate. It was stated that in many parts of the country, especially in the north, fruits were formerly grown successfully and of good quality in gardens where they cannot be grown now; and this occurred in places sufficiently removed from manufacturing centres to be unaffected by any direct deleterious influence of smoke. But an increase of cloud, and consequent diminution of sunshine, would produce just such a result; and this increase is almost certain to have occurred, owing to the enormously increased amount of dust thrown into the atmosphere as our country has become more densely populated, and especially owing to the vast increase of our smoke-producing manufactories. It seems highly probable, therefore, that to increase the wealth of capitalist-manufacturers we are allowing the climate of our whole country to be greatly deteriorated in a way which diminishes both its productiveness and its beauty.[51]

Dickens's *Bleak House* and *Our Mutual Friend*, Ruskin's *Unto This Last*, and other nineteenth-century texts with a Romantic anticapitalist bent drew apocalyptic connections between industrial emissions, the deterioration of the atmosphere, and the effects on the land: "Blanched Sun,—blighted grass,—blinded man," Ruskin puts it in his lecture *The Storm Cloud of the Nineteenth Century*, blending forms of literal and moral unseeing, as well as drawing aerial and terrestrial realms together as reciprocally influencing and defining.[52] Wallace's approach here is compatible with such imaginings but also crucially distinct: he traces out the actual cause-effect sequences—increased population bringing increased demand for manufactured items creating an increase of particulate matter from smokestacks changing cloud cover changing crop yields—that makes "climate change" less a matter of visionary speculation and more an empirically measurable phenomenon. This is the kind of ecological plot we see Darwin narrate in *The Origin of Species*; only here the chain of unintended consequences unleashed by human action is not so much an object of curiosity as a matter of grave concern.[53] Elsewhere in the book, Wallace raises other familiar-sounding alarms about clear-cutting, the loss of planetary biodiversity, and the reckless mining and consumption of minerals and fossil-fuel resources. The point here is not to set Darwin and Wallace up as opposites but to note that identifying and tracing out the environmental (in today's parlance) implications of Victorian evolutionary thought is not simply a function of hindsight but something alive and available in the nineteenth century.

But it's also worth pointing out that Wallace, ultimately, denied the full implications of evolutionary biology for human consciousness, infamously arguing that the human mind was simply too advanced to have arisen through the workings of natural selection alone. Instead, he speculated, it must have arisen through some kind of divine intervention: "These three distinct stages of progress from the inorganic world of matter and motion up to man, point clearly to an unseen universe—to a world of spirit, to which the world of matter is altogether subordinate."[54] "I hope you have not murdered too completely your own & my child," a dismayed Darwin famously wrote him in response to such speculations.[55] To Darwin, Wallace's dabbling in the supernatural threatened the integrity of the entire project, undermining his determined, lifelong efforts to rid science of all explanatory appeals to final causes or metaphysical agencies. And yet, as I hope is by now clear, we may see Wallace's swerve back into metaphysics less as a shocking act of infanticide and more as one strategy among a great many, all seeking to maintain hold on some ground for human exceptionalism against the science that so obviously threatened to undo it altogether.

MAN IN THE MIRROR

One of the elements of Darwin's theory that made it both so extraordinary and so hard to stomach was its rejection of the idea that there was any goal or purpose in nature; instead, change was ceaseless, undirected, and open-ended. There is a tension between teleology and ecology that swirled around natural selection in the decades following the publication of *The Origin of Species*. If Darwin is right, went one popular line of criticism, then why are "primitive" organisms still around today? Why haven't mollusks (to take the most common example) "evolved" yet? As an 1866 article from *British and Foreign Evangelical Review* puts it, "We are told that 'natural selection' everywhere exerts its influence, and is continually elevating in the systematic scale the different members of the organic world. There is no provision in this development theory for the persistence of low forms."[56] The confusion, as Alvar Ellegård points out, has everything to do with the metaphorical complex that categorized life-forms along a vertical axis, as "lower," "higher," "elevated," and so on, which made what was meant to be a neutral description of degree of complexity seem like an evaluative statement of intrinsic worth: "One replaced 'survival of the fittest,' which meant only the survival of those races which, for whatever reason, were able to produce most offspring able to survive to a mature age in a certain environment, by 'survival of the best,' which meant the survival of those forms which, from some arbitrarily chosen, often anthropomorphic point of view, were considered worthy of such an epithet."[57] This is obviously a profound misreading of the argument of *The Origin of Species*, but it was a common one that got taken up both by Darwin's religious-minded opponents and by "Darwinians" who were uneasy with or opposed to the stochastic, chance-driven picture of nature implied by natural selection. Ernst Haeckel, for example, one of Darwin's most influential proponents, thought that "lower" forms of life could be regarded as "living fossils" and that the great apes were, as Arthur Keith puts it, "abortive attempts at man-production."[58] As Ellegård remarks, "Such a misunderstanding, implying a confusion of values with facts, was quite common, and its popularity went on increasing: the whole of the movement of Social Darwinism may be said to be based on it."[59] But it is also a telling misreading insofar as it assumes there is a point of evolution, a final cause directing its shape, rather than a system of open ecological interactivity in which organisms all across the scale of complexity coexist in ever-changing configurations of competition and interdependence. Haeckel may have coined the word "ecology," but his conceptualization of the movement of evolutionary change was decidedly *non*-ecological insofar as it imagined a linear trajectory.[60] It is not just that such an orderly abstract plan organized around a final cause downplays the materiality of the natural world; it ultimately threatens to render that materiality irrelevant.

As Peter Bowler has argued, the "Darwinian Revolution" was not as sweeping a phenomenon as the name suggests if we consider how many of Darwin's immediate followers promoted avowedly teleological visions of evolution. This was the case with so-called neo-Lamarckian evolutionists in the latter decades of the century, like Spencer, Haeckel, Edward Drinker Cope, Samuel Butler, and a host of others, who argued that the inheritance of acquired characteristics, rather than natural selection, was the primary motor of species change; this, in turn, meant the evolutionary process could be defined by directed and purposeful activity rather than random variation and change. But random variation and undirected change form the core of Darwin's theory, the very theoretical breakthrough that finally put evolutionary speculation on firm scientific footing. The ironies abound. Although neo-Lamarckian evolutionary thought seemed to put the "environment" at the center of its analysis—arguing for the direct action of environmental conditions on the shape and behavior of organisms and thus on the course of evolutionary history—its teleological underpinnings allowed for an antimaterialist interpretation of evolution in which the natural environment mostly figures as an obstacle or opponent to be outwitted and overcome. Bowler writes: "If one chose instead to stress the act of 'discovering' a new behavior pattern to cope with a changed environment, it would be possible to see the inherent creativity of living things as a divinely implanted force that allowed them to triumph over the material world. Opponents of materialism could thus use Lamarckism as the basis for a new solution to the problem of design."[61] Others restored telos to the system in other ways. Asa Gray, the most influential defender of Darwinian thought in the United States, did not downgrade natural selection in favor of the inheritance of acquired characteristics but instead argued that it was God who generated the variations upon which natural selection was allowed to act. This was a position fundamentally at odds with the entire argument presented in *The Origin of Species*, the very power and revolutionary import of which had everything to do with the undesigned randomness of variations. As Steven Stewart-Williams puts it, "As far as Gray was concerned, the denial of intentional design in nature was tantamount to atheism. However, the denial of intentional design is the *defining feature* of Darwin's theory. Thus, Gray's perspective does not reconcile God with Darwinian evolution; it represents the rejection of Darwinian evolution."[62]

Thus, one could reject Darwin while accepting him; embrace the label and much of the terminology while holding fast to reassuring notions of design, purpose, directionality, meaning. This is at the root of the situation Gould decries in *Full House* as a "mode of special pleading," a kind of hedging that avoids having to squarely face an entirely materialist world.[63] Neo-Lamarckism and other attempts to soften or sidestep or refuse the radically decentering implications of evolution by natural selection no doubt helped maintain the centrality of the human and thus could validate what Gould terms "human arrogance."[64] A good deal of Victorian evolutionary anthropology, Lewis Petrinovich argues, was "based on the assumptions that strict Darwinism had no relevance to the understanding of social evolution. Inheritance of acquired characteristics was considered the important, and perhaps chief, cause of human social evolution, with natural selection of little importance."[65] And, more grandly, that social evolution could be understood as the capstone to the entire unfolding of life itself. As Philip Pauley notes, many scientists

> considered the inheritance of learned habits a question with great significance for understanding the relations between evolution and progress. If habits could be transformed into instincts, then evolution was presumably not merely a random material process; rather it resulted from the accumulation of experience and the growth of intelligence within living nature as a

whole. Evolution was progressive in that it represented the continual growth of Mind.[66]

The point here is that if Darwinian evolutionary discourse had broken down the boundary between humans and non-human animals, Lamarckian use-inheritance valorized culture as a way of reestablishing difference, giving humans an "out" and a way of avoiding some of the more radically decentering implications of the new biological paradigm. It's worth saying that there were genuinely open and unresolved questions about natural selection, and I'm not trying to suggest that anyone who was a Lamarckian must have been "in denial" about evolution. But just as religious-minded opponents were often quite open about the nonscientific sources of their rejection, Lamarckians like Butler and George Bernard Shaw were often willing to admit that feelings of aversion and disgust pushed them away from Darwinian natural selection and toward an alternative model. "When its whole significance dawns on you," Shaw writes, "your heart sinks into a heap of sand within you. . . . If it be no blasphemy, but a truth of science, then the stars of heaven, the showers and dew, the winter and summer, the fire and heat, the mountains and hills, may no longer be called to exalt the Lord with us by praise; their work is to modify all things by blindly starving and murdering everything that is not lucky enough to survive in the universal struggle for hogwash."[67]

Herbert Spencer, the most prominent Lamarckian in Victorian Britain, offers a slightly different and somewhat more complex version of the relationship between painful ideas and scientific knowledge. He describes the emotional "laceration" involved in coming to terms with the new scientific paradigm:

> No mental revolution can be accomplished without more or less of laceration. Be it a change of habit or conviction, it must, if the habit of conviction be strong, do violence to some of the feelings; and these of course must oppose it. For long-experienced, and therefore definite, sources of satisfaction, have to be substituted sources of satisfaction that have not been experienced, and are therefore indefinite. That which is relatively well known and real has to be given up for that which is relatively unknown and ideal. And of course, such an exchange cannot be made without a conflict involving pain. . . . The whole higher part of the nature almost of necessity takes up arms against a

charge which, by destroying the established associations of thought, seems to eradicate morality.[68]

Though he doesn't use the word, this is Spencer's account of denial: the mind rebels not against the truth claims of science but against the painful *feeling* of profound discomfort or disorientation they produce. Here is a dynamic and a structure of feeling we will see again (in very different ways) in both George Eliot and Virginia Woolf: the force to be pitted against such denialism is *self-denial*, the ability to forgo satisfactions and even deliberately embrace pain for the sake of a higher ideal like the commitment to truth and the larger cultural project. As Theodor Adorno and Max Horkheimer put it, "The history of civilization is the history of the introversion of sacrifice—in other words, the history of renunciation."[69] In this way, the pain paradoxically becomes its own consolation: it testifies to the truth of the paradigm and to the superiority of the person able to withstand it for a higher good. The body and its immediate needs can get sidelined in a process that, while potentially painful and difficult, allows it to function as an instrument of civilization. Charles Taylor writes of the profoundly "evangelical" underpinnings of this putatively antireligious position, which are also implicit in Adorno's remarks: "The turn from religion to science not only betokened a greater purity of spirit and greater manliness but also aligned them with the demands of human progress and welfare. Indeed, the courage to face the unhallowed universe could be thought of as a sacrifice which agnostics were willing to make for human betterment."[70] Believing one can divest oneself of feeling and ego in order to valiantly confront "things as they are" necessarily depends upon a strategic forgetting of one's own limits.

In other words, it depends on denying the animal body. For it is the nonhuman animal that supposedly lacks precisely the self-awareness, rationality, and willingness to abide painful sensations for the sake of long-deferred and entirely disembodied sources of satisfaction. Thus, we quickly get into a tautological knot where the very act of embracing evolutionary science and accepting human animality actually stands as a kind of proof of a "higher" and distinctly non-animal purity of spirit. The dividing line gets reinscribed in the very act of erasing it; the willingness to perform that erasure becomes part of the self-mythologizing discourse that restores it. We can see a version of this in the mini-narrative Darwin sketches at the end of *The Origin of Species*: If you bear in mind, remember, and "face" what you know is true about the natural world, there is a truth that will console you. Now the vastness of evolutionary time, by consigning *Home sapiens* to insignificance, also becomes the foundation for its cosmic singularity and importance. This was also implicit in Lyell's work, as Gillian Beer argues: "The humanistic core of Lyell's work is its insistence on the power of man's imagination, which allows him to recuperate the staggeringly extended time-scale of the physical world. Though his presence is diminished in the raw time-scale, his is the only source of powerful interpretation."[71] The diminution of the human species is accompanied by a seemingly limitless magnification of the power of human consciousness, a sense of a disembodied freedom to range across the universe. Achieving such a perspective requires a certain mind-set—the willingness and resolve to risk such vistas and extend one's mind into these dizzying reaches: "So great and grand are they," writes the physicist John Tyndall, "that in contemplation of them a certain force of character is requisite to preserve us from bewilderment."[72] Such self-aggrandizement clearly lends itself to national and racial constructions of this "force of character"—which groups possess it and which groups remain "bewildered." Thus, the willingness to grasp our lack of significance becomes the way to reestablish it; confrontation becomes avoidance, and acceptance, denial. This endlessly paradoxical situation hinges upon the human capacity for self-awareness—or, more precisely, upon the fetishized awareness of our self-awareness as the final ground of differentiation, the mark of the non-beast.

Indeed, self-conscious reflection was in many ways *the* pressure point in the conversation about human exceptionalism and the nascent field of evolutionary psychology. Opponents of Darwin were quick to press the case against his work on precisely these grounds. Consider this passage from a contemporary review of *The Descent of Man* by Darwin's one-time ally St. George Jackson Mivart:

> As to animals, we fully admit that they may possess all the first four groups of actions—that they may have, so to speak, mental images of sensible objects combined in all degrees of complexity, as governed by the laws of association. We deny to them, on the other hand, the possession of the last

two kinds of mental action. We deny them, that is, the power of reflecting on their own existence, or of inquiring into the nature of objects and their causes. We deny that they know or know themselves in knowing. In other words, we deny them reason. The possession of the presentative faculty, as above explained, in no way implies that of the reflective faculty; nor does any amount of direct operation imply the power of asking the reflective question before mentioned, as to what and why.[73]

There is something tellingly willful about this rhetoric, as if the author is not merely disagreeing with a certain conception of a fuller non-human mental life but refusing to acknowledge it and thereby almost forcibly taking it away: not "we deny that they have" certain mental qualities but "we deny *to* them" or "we deny them" those qualities. As Rick Rylance has shown, Mivart's "polemical, repetitive, furiously accusatory, and deeply out of date" attack on evolutionary psychology reveals someone facing "questions of the deepest implication and emotional conviction . . . the spectacle of a man wrestling not merely with his enemies, but with the two sides of his own convictions."[74] In his review of *The Descent of Man*, that conflict is palpable, as denial the speech act of claiming something is untrue, denial the unilateral exercise of power, and denial the mechanism of defensive refusal all sit in uncomfortably close quarters in the hammering insistence on qualitative human difference.[75]

This, as I say, is from a committed opponent. But in *Mental Evolution in Man*, which appeared about a decade after *The Descent of Man*, Darwin's protégé George Romanes partly cedes this ground to Mivart's argument in his attempt to respond to it. Although the book pursues the argument that the difference between humans and non-human animals is one of degree only, it grows unsure of itself when the subject of self-consciousness arises: "If it is possible to suggest a difference of kind between any of the levels of ideation which have now been defined, this can only be done at the last of them—or where the advent of self-consciousness enables a mind, not only to know, but to know that it knows."[76] For Romanes, human self-consciousness has slowly evolved out of more rudimentary animal consciousness in the same way that an individual human's self-consciousness emerges out of the more rudimentary awareness of infancy and early childhood; this analogy has, in other words, more than a hint of Ernst Haeckel's "ontogeny recapitulates phylogeny" argument and the developmentalist teleology encoded in it. But more than that, the "advent" of self-consciousness involves, as Romanes puts it, a "quickening,"[77] which suggests a leap into an altogether new realm of existence: "Midway between the gradual evolution of receptual ideation and the no less gradual evolution of conceptual, there appears to be a critical moment when the soul first becomes detached from the nutrient body of its parent perceptions and wakes up in the new world of a consciously individual existence."[78] Through Romanes's metaphor, continuity is transformed into discontinuity as the processes of the animal body come to birth the human nonbody. What "dawns" is not the heart-pulverizing dislocation Shaw objected to so vehemently but a new day of agency and clean separation from the rest of the world.

For Spencer, self-consciousness was the point at which a mysterious leap in human nature was effected. As he writes in *The Principles of Psychology*, "That which in the act of knowing is affected by the thing known, must itself be the substance of Mind. The substance of Mind escapes into some new form in recognising some form under which it has just existed."[79] This is indeed a kind of "escape" from the thoroughly chance-driven, materialist world, as the mysterious sudden arrival of "some new form" implicitly allows a gap between humans and non-humans to reemerge, and in that gap lies the exit from many of the radical implications of Darwinian thought. "A thing cannot be at the same instant be both subject and object of thought," Spencer argues, leading him to throw an insuperable veil of mystery over the operations of consciousness: "Mind remains unclassable and unknowable."[80] For this reason, self-consciousness was for Spencer the key to resolving any number of seeming contradictions arising out of the biologization of human consciousness: it allowed him to explain, for example, how free will (and all that proceeds from it, including the moral life) could be compatible with a biological account of the human mind and its long evolutionary development from non-human forms of consciousness. In *The Principles of Ethics*, he writes:

> Desire is a function not of consciousness, but of self-consciousness. Instead of saying that we desire the thing because it pleases, it would be nearer the truth to say that it pleases because we desire it. In not a few cases, and those not the least important, we find the value in the desire itself rather than in

any conscious gratification. In these cases we will ourselves rather than the object. All such willing is conditioned by some ideal of what we wish to be, rather than by the sole thought of something objective which we want for its own sake. It may be an ideal of vanity or of excellence, but whatever its contents, this ideal is the implicit condition and the regulative form both of the desire and of the volition. How we can form ideals and thus constitute our chief objects of desire, or how self-consciousness can modify the mechanical and passive consciousness, is beyond all telling, but none the less, is it among the most palpable facts of our inner life.[81]

Spencer is here speaking the language of neoclassical economics and, in a sense, implicitly mounting an evolutionary defense of consumer culture and the operation of the free market. Because the object of desire matters less than the form and fact of it, value is purely ideational, and the purchase of anything, including luxury goods or frivolous, disposable commodities or any of the various wasteful things social critics like Ruskin and William Morris were forever fulminating against at this time, still contributes to the person-shaping, volition-honing, ideal-forming work done by the exercise of desire itself. And in this way, real "palpable facts" that have actual material effects get produced via an intense engagement with make-believe. Indeed, read in the context of Spencer's defense of human exceptionalism, the passage is an ode to the circular value of "will[ing] ourselves" into the kind of being we "wish to be"—of a self-construction justifying itself based on the desire to be so constructed. It is telling that Spencer drills down to the magic point where continuity becomes discontinuity and self-consciousness breaks free from the "mechanical and passive consciousness" of the mere animal mind in order to "form" and "modify" and take control of things. When he reaches that crucial hinge, he finds it dissolves into mystery: "beyond all telling," as he puts it. And yet it is also "among the most palpable facts of our inner life." Although we can't explain it, we know it happens because we *know* it happens. The "reflective faculty" becomes a reflective surface, the pool of Narcissus, where self-consciousness can admire the self-confirming evidence of its own significance.[82]

Of course, this magic quickening of mechanical mental acts into free will and the formation of ideals is also, unsurprisingly, the grounds on which the "civilized" mind was distinguished from the "savage" and thus the grounds on which Europeans disavowed their otherwise undeniable kinship with other races. That is, the perceived lack of a "reflective faculty" was used to explain not only why indigenous peoples were less technologically advanced than Europeans—without self-consciousness, there is no abstract thought; without abstract thought, there is no science; without science, humans are at the mercy of their immediate circumstances, imprisoned and defined by the daily demands of survival and subsistence—but also, it was suggested, why they were less human. In a kind of implicit metalepsis, the boundaries of the discursive category "human" depended upon the extent to which actually existing humans had managed to differentiate themselves—culturally, socially, technologically—from mere materiality. Note that, in Romanes's account, the process by which difference of degree suddenly flashes into difference of kind involves an awakening that is also a "worlding"—what Ranjana Khanna calls the act of "opening up the world so that one can imagine a way of being in it"—across multiple levels of figuration.[83] Self-consciousness creates its own world of inner experience, while it also makes possible the transformation of the actual material world into an open field available to material and discursive manipulation. It is hard not to hear in "new world" a hint of the rhetoric of empire and the way the fetishization of individuated self-awareness in the West licensed the conquest of those not imagined to possess it. "In the savage," Spencer writes in the second volume of *The Principles of Psychology*, "there exists no such conception of his consciousness as that which is familiar to the civilized," which he argues is both cause and effect of an "incomplete differentiation of consciousness from material existence."[84] Walter Bagehot's *Physics and Politics*, a popular treatise on what would come to be called Social Darwinism, follows a similar line: "Every educated man has a large inward supply of ideas to which he can retire, and in which he can escape from or alleviate unpleasant outward objects. But a savage or a child has no such resource. The external movements before it are its very life; it lives by what it sees and hears."[85] Here as elsewhere in this book, the European mind is imagined as an "escape" from the merely material—it is an inward alternative to and refuge from nature's immediate pressures and demands. It is also, tellingly, its own fully stocked and supplied internal warehouse. The idea of mental "capacity" thus does double duty as a spatial metaphor setting racialized boundaries on the potential for improvement. Although the imperial project quite obviously operated according to a ruthlessly extractivist logic, the educated European comes not to *take* "resources" but to *bring* them. Indeed, he brings them with him wherever he goes! The educated, "accomplished" man, Bagehot writes, "is charged with stored virtue," as if his very body is some kind of walking battery or self-contained fuel supply, which sets up the funhouse-mirror reversal of terms in which the kinds of grotesquely unequal resource exchange that defined the imperial project could be rationalized.[86] As Frédéric Neyrat writes, "The Western relation is first of all a reversed relation. . . . The *will to exportation* serves only to reverse the origin of Western industrialization, its moment of 'primitive accumulation.'"[87] If self-consciousness was imagined as a kind of inner resource and means of transcending immediate circumstances and needs, then it represented sign, cause, and effect of European superiority, all rolled into one, conveniently, necessarily, obscuring the actual material realities of the situation.

For Bagehot, the contrast between colonized and colonizer is thus not simply one of surface and depth but also one of emptiness and fullness; of lack and abundance, all made possible through a Lamarckian inheritance mechanism that allows for the strategic, intergenerational accumulation of material experience transformed into cognitive resources. Those resources, in turn, can then be imagined as having a material value and efficacy: energy of character and force of will are energies and forces that perform work upon the world. As Spencer writes, "[Man's] circumstances are ever altering; and he is ever adapting himself to them. Between the naked houseless savage, and the Shakespeares and Newtons of a civilized state, lie unnumbered degrees of difference."[88] Here is the Achilles' Paradox of imperial ideology: The accumulation of "unnumbered degrees of difference" produces something approaching an infinite difference, as cultural achievements become encoded in the genetic substructure of the members of that culture, which then produces further cultural achievements, which piles up the differences at an accelerating rate, making them even harder to count. And thus does difference in degree blur into difference in kind. In this conception, the ability to reflect and adapt is more bank account than warehouse, with the wealth of characteristics differentiating Europeans from non-Europeans (and humans from non-human animals) accruing at compound interest. As Rylance writes about

Spencer's Lamarckian psychology, "The evolutionary process is directional, and is capable of some control by the individual whose moral or practical gains—or losses—need not end at death. Furthermore, this more 'humanized' biological process did not necessarily entail the moral bankruptcy that was claimed for evolutionary theory by anxious commentators. Putting the Lamarckian argument in terms of monetary metaphors does reveal the theoretical homology between it and some distinctive nineteenth-century theories of banking and money circulation, and Smilesian ideas of social and economic advancement."[89] We've escaped natural selection and Malthusian population pressures not simply by effacing the distinction between natural and cultural "resources" but also by translating the logic of exponential growth—notoriously frightening when imagined as a question of growing populations and resource consumption—into the more abstract, make-believe terms of cultural value. In this way, evolutionary science *itself* represents another level of reflection into which human consciousness quickens and becomes part of the upwardly accelerating process it describes. Though it takes discipline and self-denial to investigate and "face" this new picture of the natural world and the forces that structure it, to do so makes it possible to (self-consciously) harness and direct those forces to human ends in any variety of ways.[90] Thus were the imperial and the ecological in some ways antagonistically related, as the industrial domination of nature seemed to establish a hierarchy of races and cultures, which thereby justified further domination.

Darwin himself, it should be said, was ambivalent about the question of self-consciousness. His chief aim in *The Descent of Man* is to systematically dismantle the various arguments for human exceptionalism by dismantling the distinctions between humans and non-human animals. Animals, he argues, have emotions, including compassion and empathy; they can reason in limited ways; they can use tools; they even seem to have basic forms of language. What they don't have, however, is an ability to stand back and think about how they are feeling and reasoning and speaking beings: "It may be freely admitted that no animal is self-conscious, if by this term it is implied, that he reflects on such points, as whence he comes or whither he will go, or what is life and death, and so forth."[91] And yet, having said that, he immediately follows this up with a caveat: "But how can we feel sure that an old dog with an excellent memory and some power of imagination, as shewn by his dreams, never reflects on his past pleasures or pains in the chase? And this would be a form of self-consciousness."[92] Darwin is on tricky ground here—speculating about the dream life of dogs, positing the existence of a phenomenon by pointing out that you can't say for sure it doesn't exist. Embedding self-consciousness in dreams also ignores what would have been, for many of his contemporaries, its most salient quality: its ability to roam freely, actively, agentially over the conditions in which it finds itself. As Elizabeth Knoll notes about this passage, "As if aware of how tenuous the connection sounds, Darwin adds a leveling comparison," pivoting to a far more problematic—indeed, insidious—kind of speculation.[93] He writes: "How little can the hard-worked wife of a degraded Australian savage, who uses very few abstract words, and cannot count above four, exert her self-consciousness, or reflect on the nature of her own existence."[94] This is less a blurring of the boundaries than a muddying of the waters, a strategy Darwin recurs to with infamous frequency in *The Descent of Man*. If "raising" the animal doesn't narrow the gap enough, use a racist caricature of the indigenous other to cut the human (some humans) down to size. This, of course, is a way of disavowing—not denying—the humanity of non-Europeans, by which I mean simultaneously including them in category of the "human" for the biological purpose of defining the human as animal *and* excluding them from it for the anthropological purpose of defining *them specifically* as animal. As Zakiyyah Iman Jackson shows, the operative logic of racism is not a straightforward process of exclusion or denied humanity, as it is often formulated, but rather a more complex both-in-and-out, or brought-in-to-be-cast-out, configuration. As she puts it, "Black people have been *selectively incorporated* into the liberal humanist project. Blackness has been central to, rather than excluded from, liberal humanism: the black body is an essential index for the calculation of degree of humanity and the measure of human progress."[95]

We can see how this works in Darwin's own highly self-conscious articulation of the question of both denied and disavowed kinship in *The Descent of Man*. He begins by singling out what we might call "hard" denial as an impediment to the full acceptance of his theory: "Unless we wilfully close our eyes, we may, with our present knowledge, approximately recognise our parentage; nor need we feel ashamed of it."[96] To avoid being like the deniers, we must be willing to overcome whatever feelings of shame we might have about human animality. As in *The Origin of Species*, Darwin recognizes the emotional forces of desire and aversion, longing and shame, that structure what we allow ourselves to know and counsels us to confront them. So far so good. But then note that, at the very end of the book, these questions of shame, denial, and kinship return in a very different arrangement:

> He who has seen a savage in his native land will not feel much shame, if forced to acknowledge that the blood of some more humble creature flows in his veins. For my own part I would as soon be descended from that heroic little monkey, who braved his dreaded enemy in order to save the life of his keeper, or from that old baboon, who descending from the mountains, carried away in triumph his young comrade from a crowd of astonished dogs—as from a savage who delights to torture his enemies, offers up bloody sacrifices, practices infanticide without remorse, treats his wives like slaves, knows no decency, and is haunted by the grossest superstitions.[97]

Once again, the refusal of one form of denialism lays the groundwork for another. Here, Darwin is again asking us to get past our shame about being animals, but this time by leveraging the potential shame of being related to "savages," which he takes to be the harder problem. Just don't think about being related to them, he says; think about being related to animals (and, when you do, focus on their good qualities). The construction "I would as soon be descended" illustrates the act of wishful imaginative identification he would have the reader perform. It's a bizarre move, since, as the book has made plain, one does not actually get to choose one's ancestors but must simply "recognise" the hard genealogical facts. So the strategy here is not denial but disavowal: the self-conscious construction of a fiction of heroic parentage that one *knows* is fictional but chooses to believe in anyway. Indeed, there are multiple, interlocking levels to the fiction making here. Darwin's wishful genealogy, he tells us, is built upon an experience he had during his voyage to Tierra del Fuego, where, seeing indigenous Fuegians on the shore, "the reflection at once rushed into my mind—such were our ancestors."[98] Here of course he is engaging in what Johannes Fabian famously termed "the denial of coevalness"—the idea that so-called primitive peoples in the Global South are somehow prehistoric relics who do not live in the "now" of modernity. As Fabian writes, anthropology is an "aporetic enterprise" because ethnographers must simultaneously know and not know the falsehood they are perpetrating:

> Ethnographers . . . have always acknowledged coevalness as a condition without which hardly anything could ever be learned about another culture. Some have struggled consciously with the categories our discourse uses to remove other peoples from *our* Time. . . . But when it comes to producing anthropological discourse in the forms of description, analysis, and theoretical conclusions, the same ethnographers will often forget or disavow their experiences of coevalness with the people they studied.[99]

As Darwin put it in his *Journal of Researches*, "Viewing such men, one can hardly make oneself believe that they are fellow-creatures and inhabitants of the same world."[100] Easier, then, not to make oneself believe it but instead cast indigenous people back into the primitive past. Darwin can entertain the fiction that the people he sees onshore (and later gets to know personally) are his ancestors in order to refuse their equal claim on "*our* Time," but then he can also refuse to imagine that these men are his ancestors—because they aren't!—in order to dream up a bloodline more imaginatively palatable to himself and his readership. The aporetic door swings both ways, and Darwin here illustrates what Jackson calls the "plasticity" of the non-Western Other in Western discourse, "cast as sub, supra, and human *simultaneously*."[101] It's worth noting here the mixed language of agency Darwin uses in phrases like "the reflection at once rushed into my mind." The denial of coevalness is not something he imposes willfully upon these men but an almost involuntary response brought about by the shocking appearance of the Fuegians. Look what you made me do. But through the reflection *on* this reflection, he uses the operations of disavowal to exert a kind of formal management on the scene through which he can structure his feelings about it.

It's worth noting that this fetishization of self-consciousness neither begins nor ends with Darwin and his circle.[102] T. H. Huxley's grandson, Julian, who, like his grandfather, was a popularizer and advocate of Darwinian ideas, argued that the human could be freed from the story of mere reproduction and extinction, the "oscillation between impending apocalypse and reassurance of perpetual renewal," as Priscilla Wald puts it, through the offices of self-consciousness and the human ability to seize control of the evolutionary process.[103] Wald writes, quoting Huxley: "Against this oscillation, Huxley worked to tell a more agentive story of human destiny. The ability to reflect on the concept of evolution, he argued, distinguished humankind from other living organisms and came with the ethical 'responsibility and destiny—to be an agent for the rest of the world in the job of realizing its inherent potentialities as fully as possible.'"[104] Where his grandfather spoke the language of warfare and empire, Julian speaks the language of corporate management and cybernetics, calling humanity the "trustee of evolution" and "managing director of the biggest business of all, the business of evolution."[105] As Wald notes, although his interests were explicitly antiracist, Huxley's investment in the self-conscious administration of the species and therefore the world led him straight to eugenics, which he believed was not incompatible with antiracism.[106] Huxley's work has a distinctly Victorian ring to it, both in its rhetoric and its interest in the logic by which facing the humbling of humankind leads to its aggrandizement: "To me, it is an exciting fact that man, after he appeared to have been dethroned from his supremacy, demoted from his central position in the universe to the status of an insignificant inhabitant of a small outlying planet of one among millions of stars, has now become reinstated in a key position, one of the rare spearheads or torchbearers, or trustees—choose your metaphor according to taste!—of advance in the cosmic process of evolution."[107] Huxley's transhumanist philosophy lives on in the halls and open-concept workspaces of Silicon Valley entrepreneurs, many of whom continue to claim for themselves the right to direct the evolutionary course of the human species through genetic enhancement, xenotransplantation, nanomedicine, brain-machine interfaces, eugenicist strategies of population management, and other bids to upgrade the human on both individual and species levels. Much of it seems based on the loathing and denial of the animal body and the desire, as Žižek says, to "bypass material reality," even as its procedures depend upon a thoroughly materialist understanding of the human as animal.[108] As Zahi Zalloua argues, the transhumanist notion that "we are all potentially alterable" is really just a way of saying that "no one is *a priori* immune from the reach of global capitalism—and again we clearly do not *lack* immunity in the same way."[109] We will return to this in the concluding section of the book.

FREUD, EMPIRE, AND THE HUMAN

One devoted follower of Darwin, Sigmund Freud, both diagnoses and performs the kind of denialism I've been tracing in this chapter.[110] Although, as I've been trying to show, a vocabulary of denying and admitting, loosely understood as psychological strategies, circulated in the Victorian period, it was Freud who theorized both "denial" and "disavowal" as complex defense mechanisms describing, as Anna Freud puts it, "the ego's struggles against painful or unendurable ideas or affects."[111] Sorting out the ways Freud, his translators, his disciples, and his commentators have defined and differentiated "denial" and "disavowal" would require its own chapter (at least), but it's worth at least noting here the broad terminological questions at issue. In some cases, the Freudian "denial" means something like "repression," where certain facts are too aversive even to be acknowledged and thus are exiled to the unconscious, where they remain unavailable to the mind except through the work of analysis. This sense of the term of course still lingers today. "Disavowal," in contrast, is akin to fetishism (though Freud is careful to point out that the former does not necessarily involve the latter), where an unpalatable reality is accepted and rejected at one and the same time. As Freud writes in his late work *An Outline of Psychoanalysis*, "Disavowal is always supplemented by an acknowledgement; two contrary and independent attitudes always arise."[112] But the distinction between denial and disavowal is not consistently observed, in part because Freud's own ideas and definitions kept changing over time and in part because different translators have made different choices about how to render the German *Verleugnung* (which, for example, shows up as "denial" in the *Collected Papers* but as "disavowal" in the *Standard Edition*).[113] One can see the difficulty in action when, in an article dedicated to sorting out this terminology, the Freudian scholars Eugene Trunnell and William Holt cannot help getting tangled in it, producing this sentence that I had to read three or four times: "In this paper, the terms denial and disavowal are used synonymously to apply to the phenomenon described by Freud in 1927. The term 'denial' (always with quotation marks) will be used to refer to other phenomena which have subsequently been called denial."[114] In some ways, the confusion between disavowal, denial, and denial-with-quotation-marks is useful, since neither the terms nor the states they describe are fixed conditions but complex, fraught, ongoing negotiations with, for lack of a better word, "reality."

Freud's theorization of denial is inextricably bound up with nineteenth-century evolutionary thought, which means it is inextricably bound up with empire. Taking his cues from Victorian anthropology (in *Totem and Taboo* he approvingly cites Edward Tylor, John Lubbock, Herbert Spencer, and a host of others), Freud imagines the history of the entire human species as following the developmental trajectory of the individual—from childhood to adulthood, from a crude state of magical thinking to the mature capacity to face reality.[115] Emerging out of a state of denial at the social scale has, for Freud, everything to do with the changing human relationship to the natural environment, where "primitive" superstition characterized by the fear of, and subjection to, nature gives way to a more rational conceptualization of the workings of its so-called laws. This includes a thoroughly despiritualized understanding of cause and effect, both as a question of the properties and processes of the material universe and of the animal instincts motivating human behavior. As Sally Swartz notes, "There is little in Freud's theory untouched by colonialism," and clearly, such a hierarchical model of the stages of human development is an expression of imperial ideology.[116] Freud's translation of imperial hierarchies into psychoanalytic categories in *Totem and Taboo* is, as the anthropologist Donald Nonini argues, characterized by its own kind of denial—the erasure of any trace of the historical or political dimensions of empire itself from his text: "not a word about colonialism, imperialism, colonies, or empires."[117] For this reason, Nonini argues, Freud's account of indigenous peoples evinces a "studied neglect," a "forceful ignorance," and a "luxurious ignorance of a specifically informed and structured kind," because it is only through such a denial that a mythopoetic narrative of development from savagery to civilization could be constructed.[118] Nonini finds in Freud what Anne McClintock calls the "disavowed relations between psychoanalysis and socio-economic history." As she puts it, "It was precisely during the era of high imperialism that the disciplines of psychoanalysis and social history diverged," resulting in a kind of "disciplinary quarantine" that has, until recently, relegated the former to the private and the latter to the public sphere.[119] Ranjana Khanna, Zachary Samalin, and others have done much to refuse this separation and reveal the manifold links between the development of the psychoanalytic project and Western imperialism.

It is in his two late works, *The Future of an Illusion* (1927) and *Civilization and Its Discontents* (1930), where Freud theorizes denial less as a matter of individual psychology and more as a widespread cultural condition. In the former, he locates the origin of religious belief in "human weakness and helplessness" in the face of Darwinian nature, shifting the emphasis away from the father-son relationship that had loomed so large in previous work to the psychological need to defend "oneself against the crushingly superior force of nature."[120] This is denial in its "hard" form—the outright refusal of reality for the sake of a reassurance and fantasy: "[Religion] comprises a system of wishful illusions together with a disavowal of reality, such as we find in an isolated form nowhere else but in amentia, in a state of blissful hallucinatory confusion."[121] This passage (taken from James Strachey's *Standard Edition* translation) illustrates the terminological confusion—"disavowal" in this case does not signify a state of knowing and not knowing at the same time but rather a much more complete flight into make-believe (Richter, in a more recent translation, changed this to "denial of reality").[122]

Freud sees civilization as that which makes possible the progressive movement past this state of hallucination and toward a demystified, scientific conception of natural phenomena. Such a conception, crucially, privileges instrumental rationality as the means by which nature is to be confronted and, indeed, subdued: "Human civilization, by which I mean all those respects in which human life has raised itself above its animal status and differs from the life of beasts . . . includes . . . all the knowledge and capacity that men have acquired in order to control the forces of nature and extract its wealth."[123] For Freud, this marks a sea change in the human attitude toward the natural world, of course, but in another sense it traces a deeper continuity insofar as the notion of needing a *defense* against nature is still paramount. The question is where it is to be mounted. The fantastical myths that once effected the "humanization of nature"—that anthropomorphized "its dreaded forces, to get into a relation with them and finally to influence them"—can be dispensed with in favor of a *material* humanization in which those forces can be brought to heel under the auspices of industrial technology.[124] Amy Allen writes: "Freud's official position not only champions the Enlightenment virtues of rationality and autonomy; it also articulates a developmentalist story that links the ego's progressive mastery of the archaic forces of the id to the advance of civilization from 'primitive' animism through religion to scientific secularism."[125] Jemma Deer has persuasively argued that Freud's attitude toward animistic beliefs was more ambivalent and nuanced than is often acknowledged; nevertheless, we can also plainly see in much of his writing an openly hostile and reductive position taken toward complex belief systems that describe important alternative ways of understanding the enmeshment of the human within the non-human.[126] As Bruno Latour writes, "Although the official philosophy of science takes the second movement of deanimation as the only important and rational one, the opposite is true: animation is the essential phenomenon; and deanimation is the superficial, auxiliary, polemical, and often defensive phenomenon."[127] Inasmuch as Freud pursues this line of thinking, it represents a profound narrowing of the ecological imagination and a defensive posture that is internal to his own theorization of denial and disavowal. Indeed, it raises the question of the extent to which the materialized structures of "defense" (technological, infrastructural, architectural, medical) through which civilization has effected a sense of profound separation from the natural world can be differentiated from defense *mechanisms*, and to what extent the former only serves to concretize and fuel the latter.

These questions and contradictions are most clearly on display in *Civilization and Its Discontents*, where Freud touts a rising extractivist regime ("the mineral wealth below ground is assiduously brought to the surface and fashioned into the required implements and utensils") and applauds the wholesale elimination of wildlife: "Wild and dangerous animals have been exterminated, and the breeding of domesticated animals flourishes."[128] In places, he voices Promethean ambitions that would not sound out of place in the mouth of a young Victor Frankenstein:

> These things that, by his science and technology, man has brought about on this earth, on which he first appeared as a feeble animal organism and on which each individual of his species must once more make its entry ("oh inch of nature") as a helpless suckling—these things do not only sound like a fairy tale, they are an actual fulfilment of every—or of almost every—fairy-tale

wish. All these assets he may lay claim to as his cultural acquisition. Long ago he formed an ideal conception of omnipotence and omniscience which he embodied in his gods. To these gods he attributed everything that seemed unattainable to his wishes, or that was forbidden to him. One may say, therefore, that these gods were cultural ideals. To-day he has come very close to the attainment of this ideal, he has almost become a god himself. . . . Future ages will bring with them new and probably unimaginably great advances in this field of civilization and will increase man's likeness to God still more.[129]

Freud of course qualifies this: "But in the interests of our investigations, we will not forget that present-day man does not feel happy in his Godlike character," but note that the discontent he raises here does not challenge this fantasy of human divinity and the domination of nature; in fact, it is *premised* upon it.[130] Given our steady march toward total control of the environment, why aren't we happy? In the process of structuring this central question, Freud dispatches one kind of fairy tale—the wish fulfillment encoded in religious metaphysics—only to replace it with another. In a familiar move, by identifying religious believers as the "real" deniers, he opens up space for a different, quieter, "softer" kind of denialism. And that denial involves a both/and dynamic, where (again, some) humans are both humbled and exalted; identified as animal and placed beyond or above it; imagined as fully part of nature and marked off as distinct from it, even appropriately hostile toward it. No longer categorically endowed with the "likeness" of God, the species is embarked upon a journey in which it will asymptotically approach that state. The developmentalist ideology is plain, insofar as the definition of the human is grounded in the desire to achieve a status that is, ultimately, unachievable: fulfillment will always be just out of reach; thus the need to keep reaching will remain an inexhaustible imperative. The wish itself *is* the fulfillment (as any good analyst will tell you). This is why the word "almost" keeps cropping up—"of almost every—fairy-tale wish," "almost a god," and so on: such language hedges the fantasy, allows it to be invoked while also keeping it at arm's length; allows, in other words, a bit of plausible deniability.

As we see in this passage and throughout both of these late texts, the definition of the human upon which the concept of "civilization" depends is itself riven by contradiction and disavowal. One of the signal achievements of this secular civilization, according to Freud, is Darwinian science and the theorization of the full animality of *Homo sapiens* and all that follows from it: the materiality of consciousness; the implausibility of the idea of the immortal soul; the species-defining power of the sex drive and other "animal" appetites and instincts over which human will and reason have much less control than we like to believe. It is for this reason that Freud famously placed Darwin, along with Copernicus (and himself), in a pantheon of revolutionary thinkers who struck decisive blows against anthropocentric thinking—or, as he puts it, our *"naïve* self-love."[131] And yet, if one of the hard-won achievements of secular civilization is the attainment of such a more clear-eyed conceptualization of the human place in the natural order, another, Freud also says, is the technological and industrial development that makes possible a new kind of exceptionalism on different grounds. Humans are cut down to size in an ontological sense, only to be "raised . . . above" their "animal status" by means of the same techno-scientific apparatus and the hard-won self-knowledge that made it possible. As Christopher Peterson says about the "Copernican cliché," as he calls it, "The rhetorical work performed by Freud's fable of gradual human decentering not only gives us an oversimplified psychohistoriography, but in so doing too easily serves the human narcissism it is meant to diagnose. . . . *We moderns are special because we know that we are not special."*[132] Freud's rhetoric of spatial hierarchies is thus decidedly ambiguous—if there is no longer any qualitative distinction to be drawn between humans and nonhuman animals, if Darwin and other evolutionary biologists had conclusively demonstrated the complex emotional, social, and intellectual lives of apes, dogs, and other non-human beings, then what does this "above" position exactly consist of? He expresses the same kind of disavowal of the animal that we also find in Darwin and Huxley, a simultaneous rejection and acceptance of an anthropocentric hierarchy.

At the same time, and not surprisingly, Freud recognizes the persistence of denial within contemporary, secular civilization, especially when it comes to reckoning with human animality: "The element of truth behind all this, which people are so ready to disavow, is that men are not gentle creatures. . . . They are, on the contrary, creatures among whose instinctual endowments is to be reckoned a powerful share of aggressiveness. . . . Who, in the face of all his experience of life and of history, will have the courage to dispute this assertion?"[133] In this case, "disavow" has its more mixed knowing/not-knowing quality, since, as Freud notes here, the reality of "animal" violence is written all over the historical record and is something "we can detect in ourselves."[134] Indeed, what is daunting is the prospect of trying to disagree with such a conclusion, given the mountain of evidence. Just as the terminology around denial is complex and shifting, its relationship to Freud's understanding of the hierarchical stages of development is also complicated. The second edition of *A Glossary of Psychoanalytic Terms and Concepts* from 1968 defines it as "a defense mechanism of a primitive or early variety in which the ego avoids becoming aware of some painful aspect of reality,"[135] while Trunnel and Holt argue that it "involves the complex selectivity of an ego that has undergone considerable development."[136] I'm less interested here in these exegetical debates than I am in pointing out that the question itself suggests an unresolved difficulty in both Freud and the generation of psychoanalysts who immediately followed him about the adaptive value of denial. It is clear that, to the extent that it is associated with the "primitive" in the works of these writers, it is pathologized; but when it appears as a defense mechanism of "complex selectivity" at work in secular Western civilization, it is not without its utility. As Trunnel and Holt put it, "A certain amount of denial or disavowal is essential in coping with numerous environmental threats against which one is helpless or can do nothing without paying much too great a price."[137] Of course, "environmental threats" here means something like "external threats" or "threats coming from the outside world" rather than the kinds of ecological crises such a phrase inevitably calls to mind today. Nevertheless, we can see in such a formulation an argument being made for the deliberate sidelining of environmental concerns constructed through the ambiguous place of denial in Freud's anthropological narrative about the development of civilization.

And that ambiguity, in turn, is rooted in Freud's uncertainty about the value of Western civilization *itself*. This is the fault line that runs throughout *Civilization and Its Discontents*, which identifies the instinct-denying operations of civilization as *both* the very root of neurosis, psychosis, and all sorts of other psychological maladies *and* as the necessary instrument through which violent forces and desires must be managed and channeled. As Herbert Marcuse puts it, "The concept of man that emerges from the Freudian theory is the most irrefutable indictment of Western civilization—and at the same time the most unshakeable defense of this civilization."[138] Nature—both internal, as in "animal instincts," and external, as in "the environment"—becomes that which cannot be controlled and that which must be controlled. Joel Whitebook, in his recent intellectual biography of Freud, calls this paradox at the heart of the psychoanalytic project an attempt "to formulate a rational theory of irrationality,"[139] and he describes it, following Hans Loewald, as a clash between Freud's "official" and "unofficial" positions. The "official" position, Whitebook writes, is fundamentally imperial: "Maturity envisages what the philosopher, psychoanalyst, and social theorist Cornelius Castoriadis calls a 'power grab,' in which the more 'advanced' strata of the psyche dominate the more 'primitive': the ego dominates the id, consciousness dominates the unconscious, realistic thinking dominates fantasy thinking, cognition dominates affect, activity dominates passivity, and the civilized part of the personality dominates the instincts."[140] This agentive, rationalist model of the self is of a piece with what Whitebook calls Freud's "truth ethic," which is an "ethics of 'avowal.' It posits a demand that I 'own' and assume 'responsibility' for my 'interior foreign territory,' that is, my unconscious-instinctual life."[141] That ownership involves coming to a fuller knowledge about the self via the work of psychoanalysis. As Freud puts it, "Our therapy works by transforming what is unconscious into what is conscious, and it works only in so far as it is in a position to effect that transformation."[142] At the same time, of course, Freud also famously subverts the very idea of "the knowing subject" fundamental to Enlightenment thinking by insisting on the opacity of the self to itself and identifying self-consciousness, insofar as it promises a clarity it can never deliver, as simply another means of self-deception.[143] So here we arrive at the paradox of denial about which Freud himself was (unofficially) aware: In the name of divesting oneself of all illusion, freeing oneself from denial, one aspires to a position of complete command and ownership of the self, the instincts, the affect. But that very belief in a fully knowing, sovereign subjecthood *is itself a form of denial*. As Freud himself has told us.

John Zilcosky describes how this works itself out in *Totem and Taboo*, which, he shows, is a text at odds with its own archive, method, and argument. In this reading, the entire book is structured by forms of disavowal insofar as the lengthy footnotes express a profound mistrust of the work of the Victorian evolutionary anthropologists upon which the main argument of the text itself entirely depends. From the footnotes, the unofficial Freud issues trenchant criticisms about the methodological flaws through which "the savage" has been constructed (including a critique of what Fabian calls "the denial of coevalness") that the official Freud of the main text does not seem to heed on his way toward a grand conclusion about the primitive origins of the Oedipal drama: "The beginnings of religion, ethics, society, and art meet in the Oedipus complex. This is in entire accord with the findings of psychoanalysis."[144] His theory is confirmed, he assures us, but *not* just because he wants that to be the case (the analyst's eyebrow rises). Zilcosky writes: "By confidently creating this story in which his own 'presentiments' come true, Freud engages in the same wishful thinking of which he accuses savages, neurotics, and anthropologists earlier in *Totem and Taboo*."[145] *Totem and Taboo* is a text at once hyperaware and blind to itself, a condition that Zilcosky argues helps push Freud, in subsequent works, toward a conception of psychoanalysis *itself* as irrational and self-deconstructing: "Psychoanalysis, because it exposes what humanity hides from itself, has 'become uncanny to many people.' More than this, psychoanalysis has also become uncanny to itself, through its 'scientific' discovery of its own unscientific savagery. This discovery, which Freud fittingly both makes and recants, is what gives psychoanalysis its impossibly self-reflexive advantage over that other fin-de-siècle science of man: anthropology."[146] This account is compelling and suggests one of the reasons psychoanalysis has such power as an instrument of critique. And yet one wonders where that "impossibly self-reflexive" capacity actually gets us, especially given the fact that, however willing Freud might have been to build a self-critiquing method of critique, he continues, long after *Totem and Taboo*, to voice his "official" position that "Enlightenment secularism represents a civilizational advance," as Allen puts it, in ways that are necessarily "implicated in Eurocentric and racist modes of developmental thinking."[147]

One way, then, to see the articulation of Darwinian ideas in the nineteenth century and beyond is not just that it is the object of so much denial and disavowal, then and now, but that it forms a necessary part of the genealogy of those very concepts. We can see Darwin, his critics, and his inheritors working through a rhetoric of admission and denial that registered the threatened boundaries of the human in the secular liberal imaginary and the ways in which the act of reinstating those boundaries might entail some kind of strategic, necessary, and perhaps "knowing" forms of defensive refusal. Darwin himself seemed to recognize the seeds of the counter-Enlightenment idea that Freud would later make foundational to his project: that activity, behavior, indeed knowledge itself, are always indissociably bound up with desire and the body. But, insofar as the Enlightenment project and its consolations still held sway, there were also compelling demands to self-consciously exile the aversive from the mind, or from discussion, or from the field of representation. Whether this exile takes place through states of temporary suspension (as in Darwin's experience at Moor Park); through strategically fictitious and contradictory formulations of the human (as in Huxley's writings); or through the open declaration of denied kinship with non-human animals (as in Mivart's supercharged linguistic policing), it all points to a pre-Freudian history of the mechanisms of psychic and cultural refusal that not only set the groundwork for Freud's theory of denial and its riven imagination of the human but can also be seen as the constitutive contradictions out of which the theory itself emerged.

THREE

DENIAL IN THE FIRST PERSON

AS JOHN PLOTZ HAS COMPELLINGLY argued, disruptions in scale are foundational to literary naturalism and its "experiments in antiexperiential abstraction" in the latter decades of the nineteenth century.[1] What makes naturalism distinctive, he argues, is not so much its pessimism but its interest in putting human stories in direct contact with the kind of "chilly" inhuman vistas (both micro- and macroscopic) and impersonal natural forces that Darwin describes in *The Origin of Species* and elsewhere. In this way, naturalism may be seen not simply as the relatively narrow domain of the usual suspects (e.g., Émile Zola, Theodore Dreiser, Frank Norris) but as a wider "set of connected ideas, modes, and techniques responding to Darwinian thought that pervasively shaped fiction between 1859 and World War I."[2] In Plotz's account, these experiments are carried out by the omniscient narrator, and it would seem that at least part of the "chill" we feel blowing through the pages of such books comes from the superior position we sometimes hold over characters who unwittingly serve as test subjects. But we might expand his scope even further to consider the way in which Victorian fiction and poetry, especially in the latter decades of the nineteenth century, featured characters who are themselves aware of the scalar ruptures and threats to human "dignity" brought about by the new scientific picture and thus work to knit seemingly incompatible realities back into some kind of coherent whole. These texts, for this reason, frequently function both as critiques and expressions of the processes of denial, since they often show characters self-consciously struggling with unsettling forms of knowledge and the forces of desire and aversion that shape their own beliefs. Moreover, in coming to some kind of resolution about it, they stand in ambiguous relation to an author's own implied attitude toward that knowledge.

Complex formal effects are thus generated at the strange territory where the non-human and the all-too-human make their competing claims on the real. Such effects, I would argue, present us with something notably distinct from so-called unreliable narration, where the narrator's various errors (in judgment, perception, affective response) are measured against a more or less stabilized perspective tacitly shared by author and reader. The unreliable narrator is of course one way of representing denial, since it is often a telltale rhetoric of rationalization or special pleading that tips us off that something unpleasant is being actively kept at bay—think of Pip's self-serving misinterpretations of the true source of his wealth in the first half of *Great Expectations*. In this chapter, in contrast, I am concerned with situations where it is precisely the idea of the "reliable" that has been thrown permanently into crisis, where every valid claim of the real seems to entail the backgrounding of another. The contradictions and evasions in first-person narration are not necessarily set against some superior and stable moral or epistemological schema but, instead, signal the genuine difficulties involved in reckoning with impersonal, non-human realities and scales from an intensely personal, situated perspective. The key to this dynamic is the fictional depiction of self-consciousness, the reflection that both promises a means of sorting out these levels of reality and tethers the subject to a necessarily partial perspective. Thus, we might say it's a question less about unreliability as error and more about unreliability as constitutive of point of view as such, insofar as a given point of view can only ever imperfectly or provisionally apprehend what, by definition, exceeds it. There is a structuring tension between the lived sensation of distinctness and centrality and the conceptual understanding of personal or species insignificance, a split between what one knows about human life and what it feels like to be alive.

Because the question of human centrality becomes, after Darwin, such a vexed topic and because a seemingly irreducible *feeling* of centrality is part of the very matrix of first-person point of view, denialism becomes both topic and formal resource for nineteenth-century writers seeking to respond to these new scientific ideas. The various paradoxes I've sketched out here become rich, fraught, contested territory for novelists and poets struggling with the implications of materialism and the radical decentering of the human species. The poet and essayist Melanie Challenger writes: "Our intuition tells us that we are not really the creature of muscle and bone that stares out from the mirror. We are the conscious thing in our heads. In this way, we don't have to believe in dualism to be under the illusion of it. We are trapped in a sensation of personal experience."[3] That entrapment, as the political philosopher John Gray writes, occurs because "the human propensity to illusion increases with conscious self-awareness"—what appears to be the way out of denialism is actually the way straight back into it.[4]

Of course (and as Challenger acknowledges), such references to "human propensities" and invocations of "our" and "we" elide the crucial fact that this illusion of being other-than-animal is also an artifact of history and a production of culture. As we've already seen, the fetishization of self-consciousness as the final line of demarcation separating humans from non-human animals, although it predates Darwin, attained a new kind of significance and intensity following the publication of *The Origin of Species*. Self-serving and circular as it may be, the reflective faculty was the grounds on which arguments for human exceptionalism could be mounted by evolutionists, many of whom saw it, in part, as the means by which the species had made (and could continue to make) scientific, cultural, and social progress and thereby further distance themselves from their non-human and non-European brethren. It thus functioned, explicitly and implicitly, as the linchpin holding together the very category of "human," even if that linchpin was itself often something of an acknowledged fiction and even if that "universal" category somehow only included a narrow, racialized, and gendered subset of *Homo sapiens*.

Of course, these issues are always complexly ideological—interarticulated with the dynamics of race, empire, and capital. Fredric Jameson describes the difficulties involved in reckoning with structural economic realities in similar terms, pointing to an analogous kind of "decentering of the consciousness of the individual subject, whom it confronts with a determination (whether of the Freudian or the political unconscious) that must necessarily be felt as extrinsic or external to conscious experience. It would be a mistake to think that anyone ever really learns to live with this ideological 'Copernican revolution.' ... The approach to the Real is at best fitful, the retreat from it into this or that form of intellectual comfort perpetual."[5] These forms of scientific and socio-political decentering are analogous (as Jameson himself indicates with his reference to the Copernican revolution): If scientific materialism has to do with coming to terms with nonhuman scales and realities, the human subject's enmeshment in the natural, Marxist materialism has to do with coming to *in*human scales and realities, the human subject's enmeshment in transpersonal systems of extraction, exploitation, and exchange. And yet, these issues are also, on a different level, entirely intertwined, insofar as whatever obstacles impede a clear-sighted reckoning with the "chilly vistas" of scientific materialism are conditioned by the various mystifications of capital, including fictions about the autonomy and agency of the subject and illusions of control of the natural world via its wholesale commodification. That is, whatever is felt to be aversive in some existential sense about Darwinian ecology is necessarily also the product of a culture that everywhere dangles reassuring notions of distinction and insulation, both individual and social. Thus, there is a concomitant fetishization of self-consciousness as the means by which these impersonal systems might be managed and even transcended, as Jameson, again, describes: "the vision of a moment in which the individual subject would be somehow fully conscious of his or her determination by class and would be able to square the circle of ideological conditioning by sheer lucidity and the taking of thought."[6] This, he writes, is a "mirage"—the never-arriving horizon—and though in this passage he is referring specifically to a problematic dynamic *within* "Marxian ideological analysis," it seems clear enough that it is fueled by a dominant liberal humanist framework that also impeded a clear-sighted reckoning with the ecological implications of Darwinian thought.

EXTINCTION REBELLION

The attempt to bring disparate scales and realities into some kind of legible and reassuring whole is, perhaps most famously, the "plot" of Tennyson's *In Memoriam* (1850), where the speaker's grief over the death of his beloved is routed through anxiety about geological time, extinction, the random violence of nature, and the evacuation of significance from ordinary social rituals and spaces. For Tennyson's speaker, scientific naturalism not only undermined traditional sources of consolation in the face of an individual death; it also seemed to deliver a fatal blow to the human image, which seemed, in turn, to demand its own forms of consolation and mourning. The speaker ultimately finds it by turning away from the visions of mindless violence and the "repetitiousness of men and flies," to borrow a line from Wallace Stevens, that haunt the middle cantos: "men the flies of latter spring, / That lay their eggs and sting and sing, / And weave their petty cells and die."[7] That turning away is of course matched by a turning toward a spiritualized evolutionary vision of "man" rising again to the pinnacle of creation. The very genre of the elegy seems to be under threat for most of the poem, but the affective force of that genre is precisely what frames the dethroning of the human from its accustomed place of superiority as a *loss*. The natural world appears variously as the repository of now-superseded pastoral conventions, a set of "blind" abstract cosmic laws, an antagonistic realm of predation and death, and, finally, what the human must overcome to realize its destiny in "an ampler day" and with "nobler ends."[8] Nature never appears as something that could *itself* be in danger, because, with apologies to Walter White, it *is* the danger. It does not need to be mourned in its own right when it seems the chief cause of mourning.

The fluctuating, inconstant vision of the natural world in the early part of the poem suggests a sense of ontological insecurity and levels a challenge to the lyric imagination to find purchase in a world of exploded certainties. By the end, however, this very sense of instability has quietly created a sense that the environment is itself insubstantial and thus malleable under the pressure of the human imagination:

> There where the long street roars, hath been
> The stillness of the central sea.
> The hills are shadows, and they flow
> From form to form, and nothing stands;
> They melt like mist, the solid lands,
> Like clouds they shape themselves and go.[9]

This is Lyellian Deep Time, a depiction of geological change on a non-human scale that anticipates H. G. Wells's vision of time travel some decades later in *The Time Machine* (1895): "The whole surface of the earth seemed changed, melting and flowing under my eyes."[10] But where Wells employs such imagery to prepare us for his thoroughly materialist narrative of biological transformation, Tennyson uses the lyrically conjured *appearance* of insubstantiality to clear a way for spiritual resolution: science makes possible an inhuman vantage point that then becomes a more-than-human vantage point, a place of imaginative fixity in a cosmos of flux:

> But in my spirit will I dwell,
> And dream my dream, and hold it true;
> For tho' my lips may breathe adieu,
> I cannot think the thing farewell.[11]

There is more than a little willfulness in Tennyson's resolve and a touch of acknowledged denialism in admitting that there is something he *cannot* (rather than *does* not) think. As John MacNeill Miller argues in his incisive discussion of decomposition and ecology in *In Memoriam*, the poem moves linguistically and thematically in the direction of a "more total and problematic denial of decay," offering a form of poetic closure that depends upon "wishful thinking."[12] Stephen Jay Gould remarks that "we are driven to view evolution's thrust as predictable and progressive in order to place a positive spin upon geology's most frightening fact—the restriction of human existence to the last sliver of earthly time."[13] Here we see Tennyson admitting the power of this drive while allowing it to prepare the ground for the reassuring sense of evolutionary meliorism with which the poem will conclude.

And yet, even as the poem nears its conclusion, Tennyson continues to emphasize how aversion shapes what he allows himself to know. In a late canto, he argues that humans are

> Not only cunning casts in clay:
> Let Science prove we are, and then
> What matters Science unto men,
> At least to me? I would not stay.

> Let him, the wiser man who springs
> Hereafter, up from childhood shape
> His action like the greater ape,
> But I was *born* to other things.[14]

The assertive tone here is belied by the obvious goalpost shifting on display in the first stanza. The gauntlet thrown down to "Science" has barely touched ground and the speaker is already backing off—go ahead and prove this, and if you do, well, then we'll say "proof" doesn't matter. The stress that falls on the word "men" in the third line suggests an assertion not so much of the human but of a specifically masculine identity. As we saw, in his late poem "The Making of Man," Tennyson is even more explicit (and openly misogynistic) about these gendered distinctions. Here, however, the masculine posturing is immediately surrendered when we find the sentence unexpectedly ends with a sudden and qualifying narrowing of scope—from "we" to "me." The speaker begins as if he's rallying the troops but just as quickly quits the field and admits he is only speaking for—and maybe *to*—himself. In the meantime, the question mark itself seems to change from a rhetorical marker to a sign of actual uncertainty.

All of this sets up the complex epistemological and ethical position staked out in the following stanza. The italicized *born* in the last line (emphasis in the original) again strikes a note of defiance, but whether that defiance is a form of ennobling self-assertion or defensive refusal is an open question. Indeed, Tennyson does not himself seem to know quite which it is. To take it a step further, he seems to be asking whether actually there is a meaningful difference. If rejecting your animal nature keeps you from behaving like an ape, then does the fiction, no matter how false, in some sense become true? By acknowledging that belief rests on the need to believe, the speaker comes very close to admitting that human exceptionalism is really only the wish humans have not to feel unexceptional. There is a kind of heroism, Tennyson suggests, in such an oppositional posture, even if he also quietly acknowledges how tenuous the distinction is between this kind of heroism and cowardice, between fight and flight. To return to Kay Anderson's argument, discussed in the previous chapter: despite all the work done establishing the full evolutionary continuity between humans and non-human animals, "the *felt* sense of categorical distinction persisted."[15] In many ways, *In Memoriam* takes as its subject not just this gap between fact and feeling but also the additional, meta-level problem of being aware of this gap and not knowing how to think or feel about *it*: to notice how one thinks according to what one wants to feel and feels according to what one allows oneself to think. The typographical stress on *born* is thus, I would argue, a signal of actual stress, of an assertive self-construction aware of the sense of need on which it is founded. The feeling of distinction Anderson describes is powerfully compelling and, for that very reason, not to be trusted.

Something of this tension can also be seen in Tennyson's "Ulysses," and though it does not contain the same kinds of overt references to Victorian science as *In Memoriam*, it still seismographically registers its effects. The threat that occupies and impels the speaker is a vision of human life robbed of any intrinsic significance or narrative structure: "As tho' to breathe were life. Life piled on life / Were all too little."[16] The lines are, on the surface, a rejection of the placid domesticity awaiting the hero in retirement; but, clearly, they are also a rejection of what that placidity would make impossible to ignore—life as mere being, daily duration, the accumulation of moments that go nowhere, build nothing. Hence Ulysses' distaste for "hoarding," which he mentions twice in the poem—first, disdaining his own subjects who "hoard, and sleep, and feed," and then later refusing such a life for himself: "vile it were / For some three suns to store and hoard myself."[17] It's worth noting that the sense of the word changes from the first instance to the second, shifting from a moral and material register (signifying his subjects' basic appetites) to something much more metaphysical (his own unused potential energy). Of course, one cannot really "hoard" life in this second way; the whole point is that it *cannot be* hoarded, only spent, one way or another. One "piles up" life by watching the pile get smaller every day. And thus, as in *In Memoriam*, the "fight" left in Ulysses, declaimed in the famous final line—"to strive, to seek, to find, and not to yield"—is also, clearly enough, a flight from this disturbing picture of reality.[18] Boldness is cowardice; and confrontation, evasion. To strive to do what? To seek what? To find what? The whole point is that such questions are unanswerable, and maybe irrelevant. Just as the speaker of *In Memoriam* verges on admitting the circular, self-supporting logic of human exceptionalism, Ulysses here nearly gives the game away by indicating that the point of having a goal is to have one. The horizon is, by definition, all form, no content; but the form depends upon the fiction that there is content. Ulysses seems, on some level, to understand the fetishistic power of his own rhetoric, even if he also seems to find himself stirred by it. He creates a fiction to give himself something in which to suspend his disbelief.[19]

Writing on *In Memoriam*, Herbert Tucker describes Tennyson's stance as "not commitment but rather an orientation toward commitment, halfway between affirmative open-mindedness and allergic negativity."[20] He is writing specifically about this stanza, from section XXXIII, in which the speaker, describing himself in the second person, seeks to establish a position that is at once beyond the tempestuous emotions of grief and despair and also untethered to any preformulated consolations of prayer and religious ritual:

O thou that after toil and storm
Mayst seem to have reached a purer air,
Whose faith has centre everywhere,
Nor cares to fix itself to form,[21]

As Tucker points out, the deliberate indeterminacy of this belief system, its bid for a hovering kind of freedom from fixed forms, is itself an ideological construct and therefore perhaps no less fixed for being obscurely articulated and riven with contradiction: "The stanza hails a free spirit, a quintessentially liberal soul not unfamiliar in our time but whose cultural coordinates you could go back and plot on the reformist and consumerist axes of Victorian political economy—one part spiritual refugee jealous of a hard-won freedom, one part shopper having a good look around. The disposition *not* to fix on a form of faith was shaping up in the early Victorian decades as an ideology unto itself, a free-form cultural formation already recognizable in Tennyson's day."[22] The non-position position is itself a position. Hence the layers of self-consciousness structuring the pronominal play—the second-person self-address suggests intimacy and distance at once, allowing for a glimpse into this intensely private configuration of emotion and belief while also hedging and qualifying ("mayst seem") from some even purer air still further out. Tucker hears in the prosodic ambiguities of this section—which have everything to do with socially inflected questions of pronunciation—a note of discontent with the prevailing liberal ideology, even as the canto also expresses that ideology: "Tennyson invites you to choke on your own tongue, your own proper Queen's English . . . and thereby puts into all but subliminally intimate question the secular state and liberal society for which *In Memoriam* made such suave apology that you now take it for granted."[23]

The quiet hall-of-mirrors effect suggested here (and elsewhere), where one reflects on one's subject position *as* a reflecting subject who reflects on that subject position, is characteristic of the kind of mirage Jameson discusses. The self-scrutiny that seems to beckon as an escape from ideology is an expression of that ideology and thus offers no escape at all. In this canto, we can see how this is connected—albeit rather subtly—to the disavowal of scientific materialism that is taken up more explicitly elsewhere in the poem:

> Leave thou thy sister when she prays,
> Her early Heaven, her happy views;
> Nor thou with shadow'd hint confuse
> A life that leads melodious days.
>
> Her faith thro' form is pure as thine,
> Her hands are quicker unto good:
> Oh, sacred be the flesh and blood
> To which she links a truth divine!
>
> See thou, that countest reason ripe
> In holding by the law within,
> Thou fail not in a world of sin,
> And ev'n for want of such a type.[24]

Like Marlow at the end of *Heart of Darkness*, the speaker judges it best to leave the Intended (indeed, in this case, his sister Emily, Hallam's own "intended") in a state of undisturbed devotion. But where Conrad clearly means for Kurtz's betrothed to stand as the embodiment of an entire culture of denial that is gravely, destructively mistaken in its objects of faith, the sister's simple piety—though condescendingly, misogynistically rendered, to be sure—is held up as an enviable state of grace with beneficent effects in the world. Tellingly ambiguous is the quasi-prayer the speaker utters in response to her actual prayer—"Oh, sacred be the flesh and blood / To which she links a truth divine!" One way to read this is as a wish for her to maintain a sense of the sacred, in keeping with the mini-plot of the canto: let her stay as she is, free from morbid imaginings of her beloved's body rotting away in a coffin somewhere. In this sense, it's not really a prayer at all but the speaker's directive to himself not to disturb her with the shadows and doubts that beset him. This reading preserves the self-exhorting imperative mood that characterizes the rest of the canto and is thus perhaps the more obvious one. But the other way to read this is as the speaker voicing his own prayer that the human body can somehow remain spiritualized. In this sense, it is a prayer about the resacralization not just of Hallam's flesh and blood but also his sister's. Her hands momentarily come alive in the poem, "quicker" to do good and "quick" with vitality. That quickness is folded in prayer, the embodied observance of religious form that links them with "the truth divine." These two readings are not distinct options; instead, the doubleness itself reflects the speaker's own internally riven condition, his doubt that generates an internal system of endless doublings and redoublings. Am I addressing God or just myself? Am I refusing to trouble my sister for her sake or because hers is the better, truer position? The speaker tells himself not to stray into sin "for want of such a type," meaning the image of Christ as moral exemplar that organizes his sister's belief. But, of course, the word "such" is referentially ambiguous because it also could suggest that *she* potentially functions as the type for him—the image of purity, constancy, and devotion that serves as living proof of the higher nature of human beings. In short, this is a prayer that is not a prayer, a form that disavows itself. In its knowing doubleness such a stanza undermines its own desire for singleness of purpose; at the same time, its refusal of "fixed" form, its lack of commitment, maintains the mirage of freedom. We can see how the reflecting subject, above it all, tethered finally to nothing, preserves the feeling of a place beyond the merely material, the merely ideological, even as its content becomes vanishingly small with each added layer.

The ambiguous reconstitution of the knowing subject in the response to the challenge brought by evolutionary science is made even more plain in a text that, at first, would seem to have little to do with *In Memoriam* specifically or with Tennyson's poetry more generally—H. G. Wells's *The*

Time Machine. Although Wells clearly eschews the avowedly spiritual forms of consolation in which Tennyson's speaker claims refuge at the close of *In Memoriam*, he too recognizes there is something unthinkable, even for a good materialist, about the implications of scientific naturalism. At the close of the Time Traveler's grim tale of degeneration, in which he journeys into the future to witness humanity drive itself into extinction and then, further onward, to see the sun burn out in the sky, the unnamed first-person narrator makes a calculated decision to forget some of what he has just heard in order to return to the world of ordinary human experience:

> I, for my own part, *cannot think* that these latter days of weak experiment, fragmentary theory, and mutual discord are indeed man's culminating time! I say, for my own part. He, I know—for the question had been discussed among us long before the Time Machine was made—thought but cheerlessly of the Advancement of Mankind, and saw in the growing pile of civilization only a foolish heaping that must inevitably fall back upon and destroy its makers in the end. If that is so, it remains for us to live as though it were not so. But to me the future is still black and blank—is a vast ignorance, lit at a few casual places by the memory of his story.[25]

The Time Traveler's narrative has revealed the ruthless way Darwinian forces shape the future, including (in the Morlock and Eloi sections) a complexly ecological view of the reciprocally defining interactions between morphology, competitive pressures, and environmental conditions. The narrator doesn't bother trying to refute the future vision or the science informing it; instead, he chooses to "live as though it were not so," to retain a sense of individual freedom and reinvest in the ordinary world by strategically pretending things are other than they are. Here, Wells frames something very much like the "double reality" Norgaard describes: the curious mismatch between what one knows and how one feels. It's unclear what motivates the narrator's choice here, whether his refusal to accept the vision of decline is political (to counteract quietism) or emotional (to ward off despair), but that is part of the point: the seam between the double reality is where the personal and the systemic, conviction and desire, pass in and out of each other in fluid and necessarily obscure ways. It is a mark of Wells's brilliance, I think, that he uses the (for us) familiar paradox of time-travel narratives to shadow forth the uncertain moral, affective, and epistemological territory created when mismatching scales, human and non-human, come into contact. We are left wondering: Is the narrator's willed denial of the catastrophe narrative precisely the kind of response that will ensure that catastrophe will come to pass? Or is refusing to understand the future as already written the only position that makes an alternative future possible? Do forms of personal meaning making constitute a retreat into the very individualist ethos that helped create the problem in the first place? Or is the problem so incalculably vast that preserving a sphere of human meaning becomes a justifiable, even *necessary*, response, no matter how denialist it also is? I would venture to say such questions are familiar to anyone who has struggled to make sense of her own complicity and agency in the Anthropocene or to define what a meaningful life should look like in the midst of an unfolding climate catastrophe.

Adam Phillips argues that, after Darwin, "the individual person, like the species of which she is a member, is going nowhere discernible (or predictable), and nowhere in particular. But this is not so much a cause for grief as an invitation to go on inventing the future. . . . [The] new-found finality of death is integral to this open-endedness."[26] The future is written (because extinction is certain), but the future is unwritten (because no divinity is writing it for us). Phillips finds a sense of new possibility in this, whereas Wells, through the evident unreliability of his first-person narrator, remains focused on the question of how "inventing" in a tangible, material sense begins to grade into "inventing" in the more fantastical or fictional sense; or, put the other way around, he's interested in the extent to which maintaining some amount of fantasy is needed if one hopes to carry out the actual work of material invention. And then, the necessary corollary: At what point do elaborately imagined futures become merely self-protective means of not dealing with the one that is actually coming? And how would we even know?

In the novel's final lines, the narrator leaves us with a very strong, very slanted reading of the story we've just heard: "I have by me, for my comfort, two strange white flowers—shrivelled now, and brown and flat and brittle—to witness that even when mind and strength had gone, gratitude and a mutual tenderness still lived on in the heart of man."[27] It's an odd lesson to take from the apocalyptic narrative we've just heard, founded on what seems like a willful misunderstanding of the stunted, creepy romance plot between the Time Traveler and his Eloi girlfriend, Weena. The question of whether the Eloi are actually still human is one the story poses, but the narrator begs. There is a mini-tradition of strong, inadequate readings like this one closing the frame of Victorian narratives: Lockwood complacently wondering "how anyone could ever imagine unquiet slumbers" for the buried protagonists on the last page of *Wuthering Heights* (1847) (it's not that hard, Lockwood!); or Alice's older sister closing the book on Wonderland by musing on the "simple sorrows" and "simple joys" of childhood when the trip down the rabbit hole has shown them to be anything but; or Laura's tidy, recapitulative moralizing in the last stanza of the endlessly enigmatic *Goblin Market* (1862). In all of these cases—*The Time Machine* included—it seems that a bit of falsification is the price that must be paid if one wants to return to the everyday world. Where *The Time Machine* differs from these earlier texts is in its suggestion that it is not transgressive human feeling, or extreme irrationality, or polysemic psychosexual spirituality that ordinary life cannot accommodate; it is basic scientific reality. It is the dwarfing vistas of Deep Time, the corporeality of human consciousness, the inevitable extinction of the species, life, the planet itself. Thus, by the end of the novel, Wells has pulled off a rather stunning reversal: the obviously fictional journey of the Time Traveler has come to seem more "real" than the world to which the narrator (and the reader) must return. It is the everyday world, not Wells's novel, that demands a willing suspension of disbelief. In this reversal, Wells brilliantly turns the State of Art/State of Nature dichotomy described by Huxley (his former teacher) into an ironic meditation on the forms of will and desire needed to prop up the wall between them.

Of course, the idea that ordinary life (including the various structures that maintain it) is a collectively agreed-upon fiction goes back at least as far as Shakespeare, who played his own games with the suspension of disbelief and the theatricality of social and political life. And the conceit that a manifestly fantastical realm might provide greater insight into "reality" than everyday experience (because, seen properly, everyday experience *is* a fantastical realm) was a rich one for the Victorians: Lewis Carroll's *Alice in Wonderland* (1865) and *Through the Looking Glass* (1872), Edwin Abbot's *Flatland* (1884), George MacDonald's *Lilith* (1895), even Charles Dickens's *A Christmas Carol* (1843) all involve journeys to dimensions that exist alongside, function as commentaries upon, and ultimately expose the artificiality of the routines and assumptions of the ordinary world. For these writers, such multidimensionality has obvious spiritual resonance if not always straightforward spiritual significance: the idea that the material realm is an illusion and the moral journey of the individual soul the true "reality" goes back to Bunyan's *Pilgrim's Progress* (1678) and other Christian allegories, as well as to Platonic philosophy where "nature [is] but a spume that plays / Upon a ghostly paradigm of things," as Yeats puts it.[28] Indeed, the ideal is what the word "real" originally signified—the contrast is not with *imaginary* but *apparent*, as Raymond Williams argues.[29] The apocalypse the Time Traveler witnesses at the end of time is arrived at through the secularization of this allegorical mode: now the ghostly paradigm operating behind things is not a world of ideal forms or a divinely underwritten drama of sin and redemption but a scheme of ceaseless, impersonal material pressures—the collision of molecules, the unfolding of geological ages, the inescapable demands of selection and survival.

For Tennyson and Wells, then, self-consciousness offers itself as a means of demarcating the self from its material circumstances while leaving open the possibility that it expresses only the force of the need to be so demarcated. This dynamic is explored in all its unstable complexity in Robert Louis Stevenson's *Dr. Jekyll and Mr. Hyde*, an extended meditation on the simultaneous acceptance and rejection of an unwanted but irrefutable kinship. Beginning in the third person, the novel's final section, "Henry Jekyll's Full Statement of the Case," shifts to the first, a move that positions Jekyll's highly self-conscious meta-commentary as the organizing end point of the text, what will at last make sense of the mysteries and confusion of everything we've seen so far. In some limited ways, it actually does that. But whatever clarity is gained regarding the events of the preceding chapters (Where did this Mr. Hyde come from? Why was Dr. Jekyll acting so strangely at this or that juncture?) is immediately undermined by the revelation of an even more profound and unsolvable mystery about the self-conscious subject doing the narrating. That is, if that reflexive "I" is presented as the agent of stabilization, it occupies that position only by explaining how it has come undone and is not to be trusted. At times, this contradiction is presented implicitly, as when the narrator tells us he won't describe the science behind his experiments and then, without explanation, leaves two mutually exclusive lessons side by side:

> First, because I have been made to learn that the doom and burthen of our life is bound for ever on man's shoulders, and when the attempt is made to cast it off, it but returns upon us with more unfamiliar and more awful pressure. Second, because, as my narrative will make, alas! too evident, my discoveries were incomplete.[30]

We are in full Tobias Fünke mode: my experiments were doomed to failure from the start; *and also*, if only I had had more time, I could have succeeded. Irving Saposnik argues that Jekyll makes an "unavoidable admission" here, one that expresses Stevenson's own sense of personal resignation to inescapable natural and social pressures.[31] But it's crucial to see that this "admission" is accompanied by denial; that, indeed, it makes denial possible insofar as it allows him to tell himself that he has made a "full" admission. And thus the belief that one can engineer an escape from one's mortal condition survives all the carefully marshaled evidence that this is a delusion. Jekyll sets two contradictory ideas next to each other as if the second somehow flows from the first. The implicit ordering logic of "first" and "second" is set directly against the actual content of the items: the power of language to confer a sense of coherent form is matched by its power to dissolve and undermine. Stevenson does not call our attention to the contradiction but leaves it hiding in plain sight.

The contradictory attitudes toward scientific experimentation capture in miniature the generic split that structures the entire final section of *Jekyll and Hyde*, written in the first person, "Henry Jekyll's Full Statement of the Case." What Jekyll twice labels his "confession" keeps swerving into the opposite—an elaborate defense and disavowal of his own actions.[32] This occurs by way of a smoothly incoherent dance of pronouns, where the "I" of the narrative sometimes indicates Jekyll ("I announced to my servants that a Mr. Hyde [whom I described] was to have full liberty and power about my house"), sometimes Hyde ("I stole through the corridors, a stranger in my own house; and coming to my room, I saw for the first time the appearance of Edward Hyde"), and, significantly, sometimes neither ("Between these two, I now felt I had to choose").[33] That is, the confession of all the ways in which "he" is guilty always sits side by side with the slyly mooted proposition that "he" is *not* the "I" who is narrating, who is therefore not at fault ("even now

I can scarce grant that I committed it").³⁴ Hyde is definitely culpable—of murder and assault and many other undescribed acts; Jekyll, meanwhile, is somewhat guilty—of complacency and self-regard and of all the various ways he has enabled and enjoyed Hyde's depredations; the mysterious first-person narrator, meanwhile, in the very act of judging both, implies that he is neither, thus refusing his complicity in the very act of purporting to confess it. Again, Stevenson doesn't call much attention to this, letting the pronouns slide quietly from one partner to the next, although there is one notable moment where the narrative seems almost to trip over itself in confusion about its ever-changing assignments. This occurs when the narrator, as if suddenly feeling the need to assert an even firmer sense of distinction, stops to self-consciously demarcate himself from Hyde: "He, I say—I cannot say, I. That child of Hell had nothing human."³⁵ The irony, which gives the whole game away, is that this moment of verbal slapstick occurs at the very juncture in the story that finds Hyde at his *most* sympathetically human—moving about the city, acutely aware of his status as a wanted man, trying to evade detection so he can procure another dose of the magic potion. Desperate and afraid as he is, he executes an elaborate plan to restore the Jekyll identity and remains, for the most part, in command of his emotions ("the creature was astute; mastered his fury with a great effort of the will.")³⁶ These are not marks of the beast but of all the "higher" human faculties pressed into the service of that most universal of motivations, survival. Thus we feel in the narrator's desire for a brighter line drawing the force of the need to shore up a division that was never tenable in the first place. Like Tennyson and Wells, Stevenson dramatizes the way the self-conscious mind beckons as a means of distinguishing itself from an aversive and guilty kinship—primarily with the "ape-like" Hyde but also with the morally tainted Jekyll—by positing itself, in some nebulous way, as the embodiment—the disembodied embodiment—of a free and more purely "human" realm that exists beyond creatureliness, necessity, fixity, death.³⁷

Thus, as we've seen elsewhere, what beckons as a means of escape leads one only further into the trap one seeks to exit. Fittingly, *Dr. Jekyll and Mr. Hyde* ends with the narrator—whoever he is—penning his so-called confession in a locked room, waiting for the inevitable transformation back into the conspicuously guilty Hyde. The self-conscious "I," in the end, is no match for the unmanageable body and, despite its best efforts, cannot stave off the fact of the hated kinship. The novel concludes: "This is my true hour of death and what is to follow concerns another than myself. Here then, as I lay down the pen and proceed to seal up my confession, I bring the life of that unhappy Henry Jekyll to an end."[38] Note the deictics "here" and "this"—as Dora Zhang writes about the final sentence of *Mrs. Dalloway*, such words can create a kind of uncanny reality effect, especially crucial to modernist narrative: "The demonstrative statement functions to insist on the presence and the reality of the pronoun's referent."[39] Stevenson gives us one last pronominal twist, with the self-conscious narrator insisting upon his presence, his intense feeling of his own reality, just as he approaches the event horizon of his own oblivion. This has a meta-fictional component, of course, as character's end coincides with novel's. But this coincident structure, in turn, suggests the paradoxical way in which self-consciousness seems to itself both as the most real-feeling thing in the world *and* a trick of language, a fabrication. Thus in two convergent ways do we arrive at the end of something that never existed in the first place. As Stephen Arata puts it, *Jekyll and Hyde* covertly enacts a crisis in realist writing alongside its more overt thematizing of a crisis in bourgeois subjectivity, raising the expectation that the narrator will "at last write himself into the kind of coherence ostensibly promised by the autobiographical form" but instead steers both self and genre into reciprocal unraveling.[40] Whereas Wells and other Victorian writers close the "frame" of their novels by finding some kind of accommodation with the ordinary world, however provisional or however deliberately falsified, Stevenson ends with the stark and unresolved juxtaposition of incompatible realities—the highly complex form of humanizing verbal demarcation that *feels* real and the silent, undifferentiated, non-human blankness that is.

EMPIRE AND ESCAPISM

Dr. Jekyll and Mr. Hyde expresses something of the dynamic we discussed in the previous chapter: how the claims made for self-consciousness as the final stronghold of the human were inextricably mixed up with belief in white European supremacy and the ideology of empire. To the extent that Hyde is coded as a racial other—and there is plenty of material for this connection—the narrating "I" in the last section can be seen as engaged in a desperate attempt to wield its capacity for reflection as a means of maintaining the boundaries of the "civilized" and thus a racially "pure" identity.[41] The disembodied neither-Jekyll-nor-Hyde narrator very clearly divides the two named identities in racial terms: "Now the hand of Henry Jekyll (as you have often remarked) was professional in shape and size: it was large, firm, white and comely. But the hand which I now saw, clearly enough, in the yellow light of a mid-London morning, lying half shut on the bed clothes, was lean, corded, knuckly, of a dusky pallor and thickly shaded with a swart growth of hair. It was the hand of Edward Hyde."[42] The white identity is embodied and disembodied at once, marked and unmarked. Indeed, Jekyll goes out of his way to tell us that his hand was often "remarked" on, a telling insistence that also hilariously conjures these men sitting around talking about how excellent their own hands are. It is singled out as a feature of Jekyll to distinguish him from Hyde, and it is treated as self-evidently superior by the narratorial presence that demarcates and hierarchizes from its position of above-it-all rationality. "The human" inheres in a state of pure judgment and reflection available only to a certain privileged subset of the species.

Such racial undercurrents in the representation of self-consciousness are there to be seen in *Jekyll and Hyde*, but they come much more explicitly to the surface in the era's most sustained meditation on the denial of Darwinian nature, Joseph Conrad's *Heart of Darkness* (1899). In Conrad, the war that Huxley would have European civilization take to nature appears, in a series of memorable images in the novel's first section, an exercise in futility and self-delusion: recall the French man-of-war pointlessly "firing into a continent";[43] the pilgrims emptying their rifles into the same hippopotamus every night; the wreckage of industrial equipment "wallowing in the grass" and "lying there on its back with its wheels in the air."[44] The "touch of insanity" that Marlow finds in these images and operations is the force of denial that helps maintain the will to dominate in the face of the overwhelming presence of that which will not be dominated.[45] "The earth seemed unearthly," Marlow says of the landscape; "we are accustomed to look upon the shackled form of a conquered monster, but there—there you could look at a thing monstrous and free."[46] The sense of breakdown and self-cancellation rides right on the surface of the language: If "the earth" seems "unearthly," then there is a problem with the word "earth," with language itself, and the ordinary conceptual and linguistic categories through which the natural world is ordinarily understood. As Jesse Oak Taylor puts it, "No sentence in the English language better encapsulates the predicament . . . in which nature becomes symptomatic of its own removal. . . . The earth seems unearthly not because it has been disturbed, disrupted, or defiled by human presence, but because it hasn't."[47] The conventions of nineteenth-century realism that are the objects of Amitav Ghosh's critique in *The Great Derangement* (more on this in the next chapter) get dismantled from the inside out, as it were. For Conrad, it's not just that realist conventions are not up to the task of representing nature but that they have actually suppressed its reality to such an extent that the only way to access it now is through strategic, ironic excursions into alternative and unreal modes of discourse and perception.[48] As Caitlin Vandertop argues, following Fredric Jameson, "The dizzying upheavals of global capitalism and the surreal imbalances of its manifestation in the world-system's peripheries" have created a situation where "it is the real itself which becomes the true marvel."[49] *Heart of Darkness* is filled with such reversals: nature personified to emphasize its frightening depersonalization; Marlow's precise, hyperintellectual vocabulary used to produce an impressionistic, emotion-saturated depiction of the landscape. They are plainly on display in this passage, which F. R. Leavis famously singled out for criticism:

> There were moments when one's past came back to one . . . but it came in the shape of an unrestful and noisy dream, remembered with wonder amongst the overwhelming realities of this strange world of plants, and water, and silence. And this stillness of life did not in the least resemble a peace. It was the stillness of an implacable force brooding over an inscrutable intention. It looked at you with a vengeful aspect.[50]

In this case, Conrad uses the deictic pronouns—"*this* strange world of plants . . . *this* stillness of life"—to create yet another palpable reversal, the unmistakable, present reality of the place sitting awkwardly with the sense that it is not real and that Marlow on some level feels he is not *there* at all. Leavis famously complained that Conrad here seemed "intent on making a virtue out of not knowing what he means."[51] I would argue, however, that the "not knowing" might be seen here as a mode of environmental affect, the confusion that results when the bare, degree-zero picture of nature ("plants, and water, and silence") meets the need to humanize, find design, and attribute motive. The reactive and obscurely overdeveloped anthropomorphism communicates the force of Marlow's desire: better an inscrutable intention than no intention at all.

Of course, what *Heart of Darkness* also plainly illustrates are the ways in which the Victorian environmental imaginary was often deeply intertwined with imperial fantasy and white supremacy. Throughout the novel, the depiction of wild nature is consistently routed through reactionary constructions of race, and vice versa. The nature that is so disconcerting to Marlow is not merely the fecund, violent, depersonalizing, category-exploding wilderness he encounters in Africa but also the more "bestial" side of human nature that the indigenous population is used to symbolize. This, of course, returns us to Worster's argument that the drive to manage and subdue the environment stemmed in part from Europeans' desire to differentiate themselves morally, intellectually, and materially from so-called savages.[52] Characteristically, Conrad exposes the bad faith of this position, showing it failing on all fronts while simultaneously reinforcing its terms and hierarchies. The force of his critique, as Chinua Achebe argues, inheres in the notion that Europeans are no better than natives, who are imagined as the repository of all that is animalistic and irrational in the human species.[53] Something similar might be said for Conrad's representation of the natural environment; he makes it clear that the story of European civilization depends upon a similarly untenable denial of natural realities that are, when seen properly, monstrous and terrifying, the stuff of nightmares. To read *Heart of Darkness* as a climate-change novel is to notice the ways in which Conrad both critically exposes and implicitly reinforces the logic of domination that is at the root of the crisis. It is to read reparatively and suspiciously at once, to see it as a text that reveals the hypocrisy and irrationality at the root of Western anthropocentric rationalism while also maintaining its exclusionary and dehumanizing biases.

Like *The Time Machine*, *Heart of Darkness* follows a touch-the-real-and-return plot, with the key difference being that, unlike Wells's narrator, Marlow works for the duration of the narrative to locate any ground of consolation or resolution to push what he has seen out of mind. As critics have long noted, the story of *Heart of Darkness* is the story of the telling of *Heart of Darkness*, of Marlow struggling to create a coherent fiction in which he can suspend his own disbelief. Whereas Tennyson's speaker and Wells's narrator ultimately find ways to cordon off those things they "cannot think," we watch again and again as Marlow tries and fails to maintain such lines throughout the narrative:

> The conquest of the earth, which mostly means the taking it away from those who have a different complexion or slightly flatter noses than ourselves, is not a pretty thing when you look into it too much. What redeems it is the idea only. An idea at the back of it; not a sentimental pretence but an idea; and an unselfish belief in the idea—something you can set up, and bow down before, and offer a sacrifice to.[54]

"Looking into it too much" is precisely what Marlow has done, witnessing as he has the rapacity at the heart of Western capitalism, as well as the impersonal vastness and power of an "unshackled" earth. The differentiating "idea" in which he wishes to place his belief is (we imagine) the system of moral principles upon which Western civilization is founded, but these are also the principles he has seen belied and betrayed, made into instruments, masks, blinders, and weapons in the Belgian Congo. The word "idea" is repeated so often and with such emphasis in this passage that it becomes more sound than sense, leading him irresistibly to the conclusion that the idea he refers to is merely the idea of there being an idea. Marlow's idea is the same thing Ulysses is striving for, seeking, and finding—form standing in for content, signifier for signified. The difference is that Conrad's intensely self-aware narrator cannot come to rest here because he cannot help being aware of the empty, fetishistic quality of his own language. So he trails off, reboots, keeps up what he on some level knows is a fruitless quest for the consoling idea. The kind of tautological reasoning that Tennyson's speaker uneasily embraces at the end of *In Memoriam* is, for Marlow, yet another version of the motivated illogic that has helped make King Leopold's "philanthropy" possible in the first place. "The circularity, the perfect closure of the whole thing is not only aesthetically but also mentally unassailable," as Edward Said puts it, also noting that "Conrad's self-consciously circular narrative forms draw attention to themselves as artificial constructions, encouraging us to sense the potential of a reality that seemed inaccessible to imperialism."[55] Here is a crystallization of the state of denial in Conrad: caught helplessly between the enclosure of perspective and the desire to escape it that is itself an expression of it.

Heart of Darkness, of course, ends with the great, anticlimactic scene of denial, of Marlow lying to Kurtz's Intended to preserve the fiction of Kurtz the man and, by extension, Western civilization the project. In a sense, we see denial in multiple forms here—there is, of course, Marlow's ongoing crisis in internal psychic management; but there is also denial in the more straightforward, legalistic sense, as mere verbal act of gainsaying. The drama of the scene inheres in the way one form gives way to the other, as we watch the endless, epic internal battle Marlow has had to wage with himself get converted back into the usual polite opacities and falsehoods of civilized discourse. Just after he delivers the lie and makes his departure, Marlow says, "It seemed to me that the house would collapse before I could escape, that the heavens would fall upon my head. But nothing happened. The heavens do not fall for such a trifle."[56] What had been an internally seismic event now reveals itself as simply another contribution to the vast, ongoing social production of denial. The scene with the Intended illustrates what Joseph Dodds, following Stanley Cohen, writes about the relationship between everyday interactions and widespread unknowing: "When societies are confronted with collective moral responsibility for mass human rights abuses they almost always enter into collective denial, structured around an unspoken compact that certain topics will not be discussed and/or are not real. These mostly unconscious alliances are created socially, through small interactions at all levels, without any one person having to consciously create it. In complexity theory this is an example of self-organization."[57] The difference here is that Marlow's participation in this is mostly *conscious*, and yet his awareness of what he is doing does nothing to prevent him from doing it. Here is the value of all his vaunted capacity for reflection and self-scrutiny—he is divided, haunted, morbidly aware, and yet, in practical terms, not really any different from the functionaries who blindly execute the business of empire. Initially, Marlow took his commission on the *Nellie* out of a restless need to escape the everyday. "I did get tired of resting," he says before famously describing his own boyish fascination with maps and the wish to "lose myself in all the glories of exploration."[58] By the end of the novel, that expansive, escapist impulse has been transformed into the knowledge of total inconsequentiality: the wish to escape permanently fused with the acute awareness that there is nowhere to go.

THE BEAST WITHOUT

All of these issues—denial, self-consciousness, and the racialized imagination of imperial territories, even the lingering sense of contrast between earth and heaven—are brought to a kind of lurid pitch in *The Island of Dr. Moreau*. Indeed, this is the novel most openly concerned with the different forms the denial of evolution could take and the way that denial was inseparable from the ideology of empire. In *The Time Machine*, the prospect of eventual extinction is a threat that a reinvestment in the everyday, the present, and the possibility of social and political change can soften, blur, and push out of mind; in *The Island of Dr. Moreau*, the problem is different—it is not the coming extinction of the human species but its material condition as mere sentient protoplasm that proves so unsettling. That is, the problem is not the far-off fate of humanity but its current condition, and therefore the feeling of entrapment and the longing for escape are that much more palpable and immediate. As Frank McConnell puts it, "As a kind of companion piece to *The Time Machine*, *Dr. Moreau* is another post-Darwinian utopia, another evolutionary fable about human society, but one that emphasizes not the possible dangers of the human *future*, but the animal, chaotic, bloody origins and hidden nature of the human *present*, of the society in which you and I live so comfortably at this very minute."[59] After returning home from his experience on the doctor's island, where he has seen Moreau's misbegotten attempts to turn non-human animals into human beings, the narrator, Prendick, finds himself estranged from an ordinary social world in which, to him, humans all appear to be beasts: "Then I would turn aside into some chapel, and even there, such was my disturbance, it seemed that the preacher gibbered Big Thinks even as the Ape Man had done; or into some library, and there the intent faces over the books seemed but patient creatures waiting for prey."[60] This is *The Descent of Man* turned waking nightmare—now that he has seen what he has seen, there is no cultural endeavor, no work of "higher" learning, no expression of piety or effort of intellect that does not seem reducible to its basis in some kind of evolutionary function. The human world not only fails to provide a refuge from the animal; it opens the wound anew every time it promises a refuge it cannot provide. The contradiction is enough to drive the narrator mad, until the divide that he had once imagined separated the human from the non-human he now tries to wedge between the human and *himself*:

> Particularly nauseous were the blank expressionless faces of people in trains and omnibuses; they seemed no more my fellow-creatures than dead bodies would be, so that I did not dare to travel unless I was assured of being alone. And even it seemed that I, too, was not a reasonable creature, but only an animal tormented with some strange disorder in its brain, that sent it to wander alone like a sheep stricken with the gid. But this is a mood that comes to me now—I thank God—more rarely. I have withdrawn myself from the confusion of cities and multitudes.[61]

The title of the final chapter, "The Man Alone," thus describes not only the self-imposed isolation of the character but also something of the wish to refuse the messy animal ties that have deprived "Man" of his former position of cosmic singularity. "The gid," I would note, is a parasitic infection that makes sheep and goats unsteady on their feet—"Man" here is thus no longer defined as the upright, bipedal thinking being, the measure of all things— part of the answer to the riddle in *Oedipus Rex* and the basis for its humanist project—but rather as a staggering, fundamentally irrational creature, sick not from plague but from the sense that its own nature is alien to itself.

Like *The Time Machine*, *The Island of Dr. Moreau* ends on an ambiguous note, with the narrator not openly denying the plain truth of the materialist, Darwinian world he has encountered firsthand but allowing himself to feel, however vaguely, that mere matter might still, somehow, be overcome elsewhere, that something "more than animal" could remain available to the human mind:

> My days I devote to reading and to experiments in chemistry, and I spend many of the clear nights in the study of astronomy. There is, though I do not know how there is or why there is, a sense of infinite peace and protection in the glittering hosts of heaven. There it must be, I think, in the vast and eternal laws of matter, and not in the daily cares and sins and troubles of men, that whatever is more than animal within us must find its solace and

its hope. I hope, or I could not live. And so, in hope and solitude, my story ends.⁶²

Like the speaker of *In Memoriam*, Prendick confesses that it is the emotional imperative of finding "solace and hope" that structures his inquiries—he too *cannot* think certain things. And, as in "Ulysses," the vagueness of the ending is the point, as the stirring rhetoric ("infinite peace and protection," "hosts of heaven," "vast and eternal," "sins," etc.) serves to reintroduce the feeling and the form of a coherent, meaningful pattern to existence, even if it also stops short of expressly claiming such a position. Chilly as the vista through the telescope may seem, it was also (it seemed to some) an abstract, lawlike, and indeed "higher" world that might be set in contrast to the picture of the tangled, extravagant, chaotic earthly scene depicted by evolutionary biology. But more than that, since the imperial frontier and its civilizing mission can no longer serve as the horizon of disavowal (as we'll see in a moment), the horizon and the hope it can generate must be repositioned to the interstellar reaches. (The fantasy of peace that affords will soon be punctured by *The War of the Worlds*.)

The narrator's desire to avoid the unsettling truths unearthed by his own narrative is a fitting end to a novel that, on the level of plot and character, in large ways and small, is all about fleeing, hiding, evading, breaking free, and other similar acts of mental and physical release. Prendick opens his narrative describing how he "escaped" in a dinghy with two other men from the shipwrecked *Lady Vain* and later explains that Moreau scandalized the London medical establishment when a mutilated dog "escaped" from his house; that event, in turn, caused the doctor to escape the strictures of that establishment and journey to the island to conduct his research. Believing at first that he is going to be vivisected, Prendick entertains an "unreasonable hope of escape" when he realizes the door to his room has been left open.⁶³ Moreau's assistant, Montgomery, spends his nights drinking himself into oblivion to escape from everything he's had to witness all day in the laboratory. And of course, the point of Moreau's experiment is, as he puts it, "to burn out all the animal"—to find a way beyond the material. The civilizing process he puts the creatures through is its own kind of prison-house for the mind and body, and the word "escape" achieves a kind of totemic significance during the Beast People's recitation of his "Law:"

> "Evil are the punishments of those who break the Law. None escape."
>
> "None escape," said the Beast Folk, glancing furtively at each other.
>
> "None, none," said the Ape Man. "None escape. See! I did a little thing, a wrong thing, once. I jabbered, jabbered, stopped talking. None could understand. I am burnt, branded in the hand. He is great. He is good!"
>
> "None escape," said the grey creature in the corner.
>
> "None escape," said the Beast People, looking askance at one another.[64]

The culture Moreau has grafted upon the Beast People is little more than a set of words they mouth because they have to; such a scene thus becomes for Prendick a funhouse-mirror version of his own culture that he cannot later unsee—people participating in organized rituals of mystification that, on some level, they know to be exactly that. But, on another level, the idea that "none escape" rebounds ironically upon the doctor himself, whose metamorphic experiments are attempts to evade the "law" of natural selection. Although Moreau is quite obviously an evolutionist, committed to the idea of species malleability and the artificiality of various taxonomic groupings and so on, he also differentiates himself from Prendick, who studied (like Wells himself) under Huxley: "You see I am differently constituted. We are on different platforms. You are a materialist."[65] Moreau identifies as "a religious man"; specifically, he is a Lamarckian, signaled by his interest in inherited memory and his commitment "to find the extreme limit of plasticity in a living shape."[66] That word "plasticity" is the giveaway—in sharp contrast to evolution by natural selection, where morphological changes occur gradually as a result of a generational process of variation, survival, and reproduction, Lamarckian use-inheritance made conceivable the notion that any part of an organism at any stage of development could be modified and passed down. As Bowler notes, one of the implications of this susceptibility is that "one could imagine growth to be a totally open-ended process."[67] For Moreau, the barriers between species are not only not fixed; they are also endlessly permeable, giving rise to seemingly unlimited combinations of creatures— the Vixen-bear, the Horse-rhinoceros, the Hyena-swine, and so on. Though

Moreau spends his days up to his elbows in viscera, his vision is ultimately transcendent and antimaterialist, shaped by the belief that there is a way to "burn out all the animal" and escape from what seems to him the pain and waste of creaturely life.[68]

The novel very clearly sets up a conflict between Prendick's materialism and Moreau's neo-Lamarckian spiritualism and just as clearly holds the latter up as horrifically wishful and wayward, blind to its own evident implications. As Prendick says of Moreau's humanized creations, "The beast would flash out upon me beyond doubt or denial," and, by the end of the novel, the entire community of Beast Folk has undergone a full-scale reversion, abandoning their two-legged postures and creeping back into the forest.[69] But more importantly, Moreau's belief in his alternative selection mechanism is entirely undone—the humanized Beast Folk, we learn, cannot pass along their traits. In the novel's most explicit reference to the divergent evolutionary paradigms at issue, Prendick is told that "they actually bore offspring, but that these generally died. There was no evidence of the inheritance of the acquired human characteristics."[70] Moreau's failed experiment is thus a blow both to the Lamarckian notion of extreme plasticity and to the very mechanism by which that plasticity would be transmitted. "None escape" the Darwinian system. But this means—and this is the odd and crucial point—that the misanthropic nervous breakdown Prendick suffers at the end of the novel arises from him having his scientific point of view *confirmed*. He writes, "I must confess I lost faith in the sanity of the world when I saw it suffering the painful disorder of this island. A blind fate, a vast pitiless mechanism, seemed to cut and shape the fabric of existence."[71] The reference to mechanism tags Moreau's island as a microcosmic Darwinian universe: in the pride of place it gave chance variation, natural selection during this period was consistently described as soullessly mechanical: blind, wasteful, and mindless. Such terms were often used to draw a sharp contrast with a Lamarckian world defined by the purposeful, conscious, and self-conscious activity that shaped individual and evolutionary history alike.

In this way, Wells pits two different forms of denialism against one another. The first, embodied by Moreau, so recoils against the idea of a materialist cosmos that he pursues a kind of twisted evolutionary resacralization of the world via Lamarckian use-inheritance. His project derealizes the present through an imagined future that will never arrive: thus, his desire for spiritual transcendence can unfold in a charnel house, and his quest for the purely rational life can proceed through eroticized forms of torture without him noticing the contradiction. Clearly enough, this is the heart of Wells's critique of the imperial project, just as it is Conrad's: The belief in the progress of civilization can rationalize any act, no matter how horrific, as either a necessary evil in pursuit of the goal or a setback that serves only to highlight the work still needed to achieve that goal. There is no such thing as disconfirming evidence. Moreau's many failures can always be referred to the learning curve, just as Kurtz's aberrant behavior only convinces his superiors that there are kinks still to be worked out in the system. What Wells does so brilliantly in *The Island of Dr. Moreau* is highlight the way the Lamarckian evolutionary paradigm and the project of empire operate in tandem, as twinned, mutually reinforcing logics of denial. In theory, the supposedly upwardly accelerating development of Western civilization was then converted into a biological inheritance, thereby somatically differentiating civilized from savage. In practice, the inheritance of acquired characteristics gave intellectual, pseudo-scientific cover for the racial hierarchies that imperial ideology required while that ideology, in turn, pushed the case for an alternative to the discomfiting, non-hierarchical picture of the natural world painted by natural selection.

And Prendick's denialism shows just how discomfiting that picture is. A committed materialist in theory, he nevertheless recoils when he experiences that world in the flesh and then spends the rest of his life trying to rid himself of the sight. When, at the end of the novel, Prendick describes the futility and waste of Moreau's entire project, it is clearly meant as a reflection on the cosmic situation of *Homo sapiens*—bundles of flesh made self-aware by a heedless and uncaring experimenter: "They stumbled in the shackle of humanity, lived in a fear that never died, fretted by a law they could not understand; their mock-human existence began in an agony, was one long eternal struggle, one long dread of Moreau—and for what? It was the wantonness that stirred me."[72] The objection to his "wantonness" of course echoes Gloucester's lament in *King Lear*: "As flies to wanton boys are we to the gods / They kill us for their sport."[73] Fittingly, this famous analogy helps introduce Wells's own: Moreau is to the Beast Folk as nature is to humankind at large. Human consciousness is not a grand step forward but a randomly generated experiment through which protein becomes self-aware. Prendick's description of Moreau's island as revealing a "vast mechanism" indicates the curious dual function that Moreau plays within the novel. He is a character who believes himself to be a conscious evolutionary agent in a Lamarckian world but who actually, and unknowingly, functions in the symbolic economy of the novel as a kind of allegory for a Darwinian natural order: pitiless, chance-driven, undirected. Moreau is the opponent of Darwinian natural selection who, by opposing it, becomes its unwitting avatar. We are, through such logic, returned to the story of Oedipus and the way the drive to evade the "law" only further entangles one in its inescapable rigor. Of course, the putative critique of imperial ambition works at least in part by leveraging prevailing racial stereotypes and tropes, routing the fear of animal embodiment through a vision of so-called savagery in the form of the Beast People. The fact that, for Prendick, their "primitive culture" holds a mirror up to his own may be usefully demystifying, but it also depends on a reduction of indigenous cultures to something self-evidently repellent and ridiculous: we're just as bad as *they* are, goes the racist account.

Through Prendick's journey and return, Wells stages a confrontation with what Wynter calls "the overrepresentation of Man"—that is, the way the inclusive biocentric category "human" is used, paradoxically, to enforce the exclusive rights to that category by a privileged white, male, European, bourgeois subset of humanity. We have seen Huxley's "Man," transfigured on a mountaintop, standing upon a similar set of open contradictions and disavowals: *from* but not *of* the natural world; occupying a "place" in nature imagined as no place; embodying the species through a rise into disembodied figuration. Moreau himself is this overrepresented Man—wreaking havoc on the world in a futile attempt to transcend it—and through witnessing the collapse of that position, and the bad faith and self-deception that had structured it all along, Prendick confronts not just the undoing of human exceptionalism but of the racial hierarchy upon which it always depended. The "dead bodies" he sees riding the trains and buses in London suggest the "necropower" that Achille Mbembe argues structures the relationship between colonizers and colonized; in his account, the former claims sovereignty and full humanity though the power to expose the bodies of the latter to mere animal existence and a kind of death-in-life. Mbembe writes: "That colonies might be ruled over in absolute lawlessness stems from the racial denial of any common bond between the conqueror and the native. In the eyes of the conqueror, *savage life* is just another form of *animal life*."[74] At the end of *The Island of Dr. Moreau*, the "common bond" that unites colonizers and colonized is a shared humanity, but not in the idealized liberal version of that phrase; the commonality inheres, instead, in the way all humans are imaginatively reduced to the status of corporealized subjects, the walking dead. No one can claim special representational status on the mountaintop; all that is left is the absolute leveling representation of humans as the despised animal "multitudes." No longer to be found in European "civilization" or its various cultural fictions and overrepresentations, the last escape for Prendick appears to him in the form of a makeshift kind of personal exceptionalism—what Timothy Morton calls "beautiful soul syndrome"—in which he indulges the feeling that he can somehow sunder the ties that bind him to the rest of his species, including his own corporealized countrymen.[75] The retreat from the social world into the observatory and the world of astronomy is his way of turning a blind eye to the human by directing his sight elsewhere: "There it must be, I think, in the vast and eternal laws of matter, and not in the daily cares and sins and troubles of men, that whatever is more than animal within us must find its solace and its hope. I hope, or I could not live."[76] As Wynter says, the overrepresentation of Man was built upon "the systemic stigmatization of the Earth in terms of its being made of a 'vile and base matter,' a matter ontologically different from that which attested to the perfection of the heavens," and Prendick here, despite all he has experienced, can still be seen desperately clinging, like a shipwrecked person, to one of the planks of a splintered human exceptionalism.[77] But I would also note that in his own elaborate self-report of his investment in the astronomical—describing the need for hope that structures what it is he thinks and feels about the heavens—he is really clutching onto the dream that the "more than animal" can be found in the operation of reflection—in the reflective faculty itself.

As many critics have noted, *The Island of Dr. Moreau* was written as a direct response to Kipling's *The Jungle Book*, with the Beast People and their mindless recitation of "the Law" meant as a send-up of Kipling's talking animals and their habit of waxing grandiloquent about the "Law of the Jungle." As Darko Suvin writes, "*The Island of Doctor Moreau* turns the imperial order of Kipling's *The Jungle Book* into a degenerative slaughterhouse, where the law loses out to bestiality."[78] Or, to modify this slightly, the "law" gets revealed as a cover for exploitation and excesses that may be called "bestial" but are well-established fixtures in the repertoire of human behaviors. *Moreau* inverts and mocks *The Jungle Book* in all sorts of ways, but what I want to emphasize here, in closing, is the fact that its critique hinges upon two different evolutionary paradigms at play and, further, that those different paradigms have significant formal implications not only for these two texts but for late Victorian narrative more generally. Kipling was and is often tagged as a "Darwinian" writer—a contemporary review of *The Jungle Book* in the *Atlantic*, for example, argues that the stories can be read as extensions of the arguments found in *The Origin of Species* and elsewhere: "Mr. Kipling, indeed, has expounded relationships in the psychology of the animal world as far-reaching as those which Darwin discovered in its morphology"—but Kipling did not disguise the fact that he *loathed* Darwin.[79] In his letter to the writer Edmonia Hill, he describes the same kind of visceral, emotional response to Darwin that we saw in Shaw: "I've been trying once more to plough through *The Descent of Man* and every fiber . . . of my body revolted against it."[80] The idea that Kipling was some kind of fabulist for the post-*Origin* evolutionary moment perfectly illustrates what the historian of science Peter Bowler has termed "the non-Darwinian Revolution" in the late nineteenth century—the idea that much evolutionary discourse that got (and gets) reflexively labeled as "Darwinian" in fact owed much more to Lamarck and kindred developmentalist theories about species transformation; such discourse in many ways was actually *anti*-Darwinian because it stood in stark opposition to the undirected and chance-driven model of growth and change that made his theory so groundbreaking in the first place. As I have shown elsewhere, Kipling makes numerous references to the inheritance of acquired characteristics as well as to the theory of recapitulation popularized by the German biologist Ernst Haeckel, in which an individual organism passes through the same stages of the evolutionary growth of the species on the path to attaining its adult form.[81] Lamarckian use-inheritance and recapitulation theory do not, strictly speaking, entail each another, but, in practice, they tended to come as a package since both present a retentive, progressive, and hierarchical alternative to the stochastic and open-ended picture of the natural world painted by natural selection. For a time, Wells had himself been a Lamarckian but rejected it in favor of the so-called neo-Darwinian August Weismann and his arguments about the "all-sufficiency of natural selection."[82] Seen in this context, then, the self-professed Lamarckian orientation of Wells's anti-hero Moreau functions as a pointed critique of Kipling's anti-Darwinian mythmaking.

Lamarckian evolution gave self-consciousness pride of place since the human mind was conceptualized as both as the highest achievement of evolutionary unfolding and as the one agency in the universe that could understand itself and thus shape the future course of its own development. As we've seen, for many Lamarckians (though not only them), the human mind is evolution becoming self-aware. In attaining that awareness, the mind also, paradoxically, steps outside the process, learns to direct and manage the future, and thus frees *Homo sapiens* from servitude to nature, necessity, and the merely "animal" instincts. Thus, in *The Jungle Book*, it is not just Kipling or the third-person narrator who traces out the evolutionary implications of the narrative arc but the character himself who comes to understand his own story in these terms. Near the conclusion, Mowgli recapitulates his growth in a way that obviously recalls Haeckel's biogenetic theory of ontogeny recapitulating phylogeny: "'Mowgli the Frog have I been,' said he to himself; 'Mowgli the Wolf have I said that I am. Now Mowgli the Ape must I be before I am Mowgli the Buck. At the end I shall be Mowgli the Man.'"[83] He realizes here both the narrative trajectory toward the top of the animal hierarchy and his status as a kind of embodied compendium of animal strengths and attributes—the animal of animals and therefore no animal at all.[84] It is not that the character simply understands his own identity as "master" of the jungle; it's that he achieves that identity *through* the understanding of it.

This, clearly enough, is also the point of contestation between the evolutionary visions of Kipling and Wells. If self-awareness in *The Jungle Book* both indicates and enacts a new kind of mastery over nature, it appears in *The Island of Dr. Moreau* as little more than a gateway to self-deception. As we have seen, both Prendick and Moreau are quite prepared to step back and reflect upon their personal narratives, their relationship to non-human animals, and the situation of humanity as material beings in a material universe. And both of them, by means of that reflection, *because* of that reflection, and the belief it engenders that they have somehow achieved some superior, hard-won insight into capital-R Reality, only get themselves more deeply mired in evasion and special pleading. Wells, at this stage of his career at least, is deeply skeptical of the power of human consciousness, either as a marker of evolutionary superiority or as a means of actually achieving anything more than the illusion of distance from the material world. Indeed, insofar as the Beast People function as a grotesque, funhouse vision of the human, the idea that ambulatory clusters of protoplasm have attained the means to consider their situation appears a horrifically cruel cosmic joke. Self-consciousness represents no evolutionary pinnacle but something of a dead-end, a failed experiment intent on convincing itself of its success. In other words, what looked to many like a means of escape from the merely material here leads only deeper into denial and thus an even more insidious kind of enclosure.

FOUR

GEORGE ELIOT AND FREE INDIRECT DENIAL

IN *THE GREAT DERANGEMENT*, Amitav Ghosh argues that the generic conventions of the nineteenth-century realist novel formally encode a kind of environmental denialism that makes phenomena like anthropogenic climate change difficult to "think." As he puts it, "The very gestures with which it conjures up reality are actually a concealment of the real."[1] Insofar as realism tends to privilege the individual life story over collective experience, the plausibly quotidian over the freakishly improbable, the socially bounded world over the non-human, Ghosh argues, it makes it more difficult to come to terms with something as transpersonal, as dynamically and complexly intermeshed, and as catastrophically discontinuous as climate change. Literary criticism has long been suspicious of the ideological baggage that gets smuggled in under realism's implicit claims to representational transparency, but Ghosh locates the pressure points in slightly different places, questioning not so much what kinds of individual lives are afforded space in the narrative world, or how they are represented, but whether the focus on what he calls "the individual moral adventure" might make it harder to grapple with environmental disasters born of, and only remediable through, collective activity.[2] Or does the basic adherence to probability, he asks, woven into the fabric of realist plotting, carry with it falsely reassuring notions about the stability of the biosphere and the perdurability of the human species? If such emphases and assumptions define the forms through which we process the world cognitively, emotionally, implicitly, then they also help define the way we process the world materially, irreversibly, actually. According to Ghosh, they explain why climate change remains so stubbornly "unthinkable," for they both reveal and maintain the impoverishment of the Western ecological imaginary.

The broad strokes with which Ghosh paints his picture of realism explains something of *The Great Derangement*'s popularity outside academic circles and mirrors his critical investment in privileging the macro over the micro, the boldly struck over the ploddingly continuous. But the lack of interest in the fine details, fitting as it might be in some respects, can make the argument seem thin on the ground in others, especially when we're talking about a form as capacious, as variegated, and as frequently self-aware as "literary realism." In this chapter, therefore, I want to consider a defining feature of the nineteenth-century realist novel—free indirect discourse—and the complex ways in which it both critiques and reproduces some of the denialist habits of mind Ghosh identifies. Roy Pascal calls free indirect discourse a "dual voice" since both narrator and character seem to speak at once: the character's distinctive idiom and point of view are transmitted without quotation marks and with the pronominal indices of the third person.[3] It becomes recognizable *as* free indirect discourse, therefore, only when we can identify the biases or misperceptions or desires rooted in an individual character's perspective and needs—things the third-person narrator obviously wouldn't say or think. This dual quality, I will argue, makes it an instrument uniquely capable of expressing divided minds, forms of self-deception, selective attention, motivated reasoning, strategic forgetfulness, and wishful thinking; that is, states of denial. In its capacity to take the silent measure of characters and expose the various verbal tricks, screens, and stratagems by which they imperfectly cloak uncomfortable truths from themselves, free indirect discourse implicitly installs the narrator as the guarantor and silent spokesperson *for* those truths. The "real" emerges in an implicitly imagined consensus understanding of reality against which any individual departure can be assessed, ironized, and corrected. And yet, because of this, because of the consensus view that the device both relies upon and constructs, free indirect discourse is particularly susceptible to becoming a means of mystification itself, an instrument (to return to Ghosh) for concealing the real through the very gestures with which it claims to uncover it.

My focus here is on George Eliot because, maybe more than any other nineteenth-century English writer, she sought to put the "individual moral adventure" in touch with wider, transpersonal realities and networks of action and interaction. As generations of critics have noted, Eliot's narratives of moral development have everything to do with the process by which characters subdue the "immense need of being something important" (as she says of one of them) and come to terms with the reality of their own obscurity and insignificance.[4] All of her works, but especially her late novels *Middlemarch* and *Daniel Deronda*, dramatize a long and painful reckoning in which a felt and, in many ways, physically embodied sense of personal centrality chafes against the discordant knowledge of a much wider, thoroughly impersonal, and increasingly "globalized" world teeming with other "equivalent centre[s] of self."[5] The multi-plot narrative structure, with its plethora of connected stories, vectors of mutual influence, and simultaneously existing centers of consciousness, has, in her hands, an obvious ethical dimension insofar as it makes unmistakable the peripheral status of even the most narratively centered points of view. But a belief in one's personal "supremacy" (to use Gwendolen Harleth's term) is not just illusory and morally bankrupt; it is also dangerous, insofar as it obscures the true conditions of one's life and therefore prevents one from making decisions that reflect the realities of those conditions. Thus, what first appears as distressing, aversive knowledge ultimately holds the key to liberation from the false comforts and sham sovereignty of selfhood. More than any other writer in the period, Eliot brings to life the variety of defenses mounted against that aversive knowledge and reveals the process through which those defenses are gradually overcome. That is to say, she brings to life the workings of denial.

Eliot's interest in denial is inseparable from her investment in realism. By showing us the motivated processes through which individuals work to disguise reality from themselves, she allows that reality to gather a kind of silent power in the negative space of the narrative. This works hand in hand with the reality effect created by free indirect discourse, where the character's distinctive voice becomes "audible" to the reader inasmuch as it registers some kind of break or departure from an unstated (because assumed) consensus understanding of reality. In this way, realism makes its claim more in the breach than in the observance. As David Carroll puts it,

> What [Eliot] demands is that we become sensitive to the exact point where the mind meets the outer world, where the hypothesis comes into contact with the facts it is trying to explain, where the deduction begins to mould the evidence.... This is the crucial area of interaction. In one sense, this *is* the reality of the novels—not the mind, not the external, but their meeting place. This is what George Eliot means when she refers repeatedly to the "medium" of Middlemarch society. It is the combination of all those intermingled webs spun between the mind and the external world, and, in the absence of any coherent social faith and order, it determines subtly and firmly the way in which life is interpreted and whether individuals are sacrificed or redeemed.[6]

The "meeting place" Carroll describes is thematized and remarked upon in various ways, but it is also *voiced*. The medium of *Middlemarch*, for example, can be understood as the manifold networked interactions that define social life in the diegetic world, but it is also the narrative language in which its disparate idioms mingle, clash, ventriloquize, ironize, and fade into and out of each other. If, as Franco Moretti says, it is an "emotional spark" that "brings free indirect style into being," Eliot allows us to hear the voltage humming through the electric fence at the place where mind and world meet.[7]

Part of my aim in this chapter is to show what a subtle and sophisticated instrument free indirect discourse is in Eliot's hands and how she uses it to render the ambiguously mixed affective and epistemological condition of denial. But I also want to show the way that the construction of the denied "reality" itself involves a different kind of denial insofar as it deliberately and, at times, explicitly excludes the non-human. One way to put this is to say that she attempts an uneasy balancing act, a needle-threading procedure whereby the narrative works everywhere to challenge the delusions borne of individual exceptionalism (variants of that word run throughout *Daniel Deronda*), while shoring up a compensatory sense of human exceptionalism. Indeed, it is the reassurance of the latter that allows her to push for the former: the surrender of a sense of individual importance is insisted upon in the name of human development, historical progress, and (finally) national and racial identity. In this way, Darwin is a double-edged sword for Eliot. On the one hand, the demystifying power of his work helps dismantle sentimental pretenses about human motives, revealing the predatory desires that often get masked and naturalized by social forms. We might recall the famous moment in *Middlemarch* where Mrs. Cadwallader's matchmaking is compared to the voracious feeding of an amoeba.[8] On the other hand, Darwin's more radically decentering ideas—the loss of qualitative distinctions between humans and non-human animals, the undirected course of species development, the inevitability of extinction—threaten to reframe the human story writ large as a chaotic scramble. As I discussed briefly in the Introduction, Gillian Beer argues that natural selection made Eliot deeply uncomfortable, since it belied the idea of intrinsic or necessary progress often mistakenly imagined as the meaning of "evolution." Beer writes: "[Natural selection] is an idea so much at variance with George Eliot's own morality that it is not surprising that she did not immediately grasp its implications. Whenever she refers directly to the idea of natural selection, that faintly facetious orotund style appears, to which she is driven by ideas that cause her deep disquiet and which she yet cannot repudiate."[9] Note that Beer's "and yet" suggests the divided state of mind we've been tracking, the uneasy status of an idea acknowledged but not fully admitted.

Beer's account is all about *voice*, about hearing the way Eliot's tone simultaneously expresses and suppresses her unease. As I hope to show, the operations of denial can also be heard through her strategic use of free indirect discourse, which functions both as an instrument that sounds denied realities and a means of concealment and suppression. The moral development of her characters depends upon their coming to terms with an existence as a mere unit within a much larger whole, but that larger whole is framed as a *human* one: a community, a race, a nation. What they do not reckon with, what they are in fact encouraged to keep out of mind, is the idea that they are mere units in a greater *non-human* whole that is far more threatening and destabilizing. Eliot, John Kucich argues, "is appalled by man's universal and irreversible amalgamation with matter," and thus she works to shore up a threatened distinction between human and non-human.[10] For Eliot, this distinction has everything to do with history; or, more precisely, it has to do with the historical imagination, the achievement of which allows individuals to surrender their own personal desires for the sake of a greater and meaningful transpersonal good. Eliot seems to acknowledge the circularity of this: human progress depends on the necessary fiction that there is such a thing. As we saw with Tennyson, humans are exceptional by dint of their ability to believe in their own exceptionalism. Thus, we approach the uncomfortable proposition that to be human is to be capable of denial. In what follows, I show the way Eliot works through these paradoxes chiefly through readings of her late novels *Middlemarch* and *Daniel Deronda* and their sophisticated use of voice. These novels not only represent the pinnacle of her artistic achievement and her fullest realization of the expressive possibilities of free indirect discourse; they also show her most fully working to manage her own response to the implications of Darwinian science.

VOICES OF DENIAL

It is no exaggeration to say that free indirect discourse is a defining feature of nineteenth-century fiction. Ian Duncan calls it "the normative technology of Victorian realism";[11] and Frances Ferguson argues that it is "the novel's one and only formal contribution to literature."[12] The device has been of longstanding interest to narratologists and formalist critics like M. M. Bakhtin, Gerard Genette, Wayne Booth, Dorrit Cohn, and those who have sought to differentiate it from other modes of representing consciousness in literature (direct discourse, internal monologue, psycho-narration, etc.), to define its grammatical and stylistic peculiarities, and to understand the various expressive possibilities created by the ambiguity of its dual voice. Partly because of the way it presses an individual character's perspective through a remote, impersonal voice without that character's awareness, free indirect discourse has often been seen as a device dragging along heavy ideological baggage. Moretti, for example, argues that it functions to bring subjects in line with social doxa: "By leaving the individual voice a certain amount of freedom, while permeating it with the impersonal stance of the narrator, free indirect style enacted that *véritable transposition de l'objectif dans le subjectif* which is indeed the substance of the socialization process."[13] Foucauldian critics like D. A. Miller and Vincent Pecora find in it a formal expression of an emergent surveillance regime, of the internalization of power structures that condition a subject's point of view.[14] As Miller puts it, "The master-voice of monologism never simply soliloquizes. It continually needs to confirm its authority by qualifying, canceling, endorsing, subsuming all the other voices it lets speak. No doubt the need stands behind the great prominence the nineteenth-century novel gives to *style indirect libre*."[15]

Whether the social world is imagined as a totalizing disciplinary system or something more open with possibility, the third-person component of free indirect discourse—at least in most nineteenth-century realist novels—is almost always discussed as the conveyor of the set of normative, socially agreed-upon values that constitute "reality." As Ferguson puts it, "The most nearly indisputable representation of a general will or social character appears in the narrator's free indirect style."[16] We hear the character's distinct voice through the sense of misalignment—through a modifier conveying a prejudice we know the narrator doesn't share:

> Gwendolen's dominant regret was that after all she had only nine louis to add to the four in her purse: these Jew dealers were so unscrupulous in taking advantage of Christians unfortunate at play![17]

. . . or through the voicing of questions we know the narrator wouldn't ask:

> He was trembling now with annoyance. Why did she seem so abstracted? He did not know how he could begin. Was she annoyed, too, about something?[18]

. . . or through the exclamation of something we know the narrator isn't feeling, or at least isn't feeling quite so intensely:

> Dean Arabin had laughed at him because he had persisted in walking ten miles through the mud instead of being conveyed in the dean's carriage; and yet, after that, he had been driven to accept the dean's charity! No one respected him. No one![19]

Free indirect discourse makes palpable the ways in which the individual self is circumscribed—by desire, by fear, by shame, by the needs of the ego, by the exigencies of the moment—and how that circumscription necessarily, definitionally, puts it in some state of tension with a broader shared reality. In recognizing the irony, in hearing the dual voice of free indirect discourse, in gauging the degree of the mismatch between a socially constituted reality and an emotionally pressurized inner realm, the reader herself is drawn to the side of, is invited to share in, the consensus position represented by the narrator.

In this way, in the ironic coordination of the dance between the demands of a putatively impersonal reality and individual "self-delusions,"[20] or "secret vanities and self-deceptions,"[21] free indirect discourse frequently traffics in the readily identifiable language of rationalization, wishful thinking, self-justification, and motivated reasoning: the language of denial. It draws our attention to what the narrator of *Middlemarch* calls "that wordy ignorance which he had supposed to be knowledge"[22]—the way in which the loaded words and phrases that ostensibly mark off a character's distinct voice from the narrator's are, at the same time, recognizably conventional terms that are being used to manage or conceal uncomfortable facts. As Elaine Freedgood argues, "Free indirect discourse is not so much a representation of subjectivity as a representation of the intertextuality of representations of subjectivity, including those we make to ourselves about ourselves."[23] The novel she has in mind here is Gustave Flaubert's *Madame Bovary*, in which the protagonist's interiority is so saturated with the generic and linguistic conventions of romance novels that her individual voice speaks through them. If, as Gilbert Chaitin argues, Flaubert "perfected" free indirect discourse, he also pushes it to a kind of extreme, willing as he is to lay bare the threadbare intertextual fabric of Emma's mind and humiliate his own protagonist.[24] What is often held up to be mocked in *Madame Bovary* and other nineteenth-century realist novels is not simply the *fact* that a character speaks in formulaic expressions but that he or she does so in order to push an otherwise uncomfortable or even intolerable reality out of awareness. Here are three very different examples from Flaubert, Hardy, and Tolstoy:

> She had been caught in it all by some accident: out beyond, there stretched as far as eye could see the immense territory of rapture and passions. . . . Didn't love, like Indian plants, require rich soils, special temperatures?[25]

> But Sue was so dear! . . . If he could only get over the sense of her sex, as she seemed to be able to do so easily of his, what a comrade she would make.[26]

> There was a little thing, a tiny little thing, in the appendix. This could all be put right. Strengthen the energy of one organ, weaken the functioning of another, absorption would take place, and all would be put right.[27]

Pressed through different idioms (the romantic, the sentimental, the medical), all three passages use free indirect discourse to highlight a process whereby language attempts to fend off a disquieting reality. For Emma Bovary, that reality is the sense that her actual life does not allow for the kind of sustained emotional intensities promised by her reading. For Jude, it is the unnerving suspicion that he is driven by his own sexual desires. For Ivan Ilyich, it is the dawning awareness that his bodily ailment is grave and irreversible. In all of these examples, denial is part of a struggle against the forces of an unvarnished reality that would undermine an individual's sense of living an autonomous or distinct existence: Emma wants to believe her life is not defined by the quotidian; Jude wants to believe he is not a captive to his biology; Ivan wants to believe he is not marked for death like everyone else. But because the third-person narrator slips into these phrases and thereby ventriloquizes a character's telltale personal idiom, the desire for a sense of a distinct, individuated self is undercut by the very deindividuating procedure of the form through which that desire finds expression. In all of these cases, free indirect discourse calls attention to *how* language is being used in a way that also reveals *why* it is being used that way. We clearly see and, in a sense, feel, the very human need to explain and to rationalize, to talk oneself into more palatable, self-sustaining, even immortalizing versions of reality. This is the paradoxical reality effect created by free indirect discourse: it is the *gap* between character and world that makes both character and world seem "real." The misperception prompts the reader to imaginatively correct it and fill in the "actual" situation. This is a fiction too, but it is in closer touch with reality than the obvious make-believe in which the character dwells. Conversely, it is through recognizing the implied, unspoken desire motivating the construction of that make-believe that creates a sense of the force of a perceiving mind in a state of ongoing tension with what lies beyond it. The energy of an inner life, vital and real to itself, emerges, counterintuitively, by its use of the conventional and the need to voice certain things for the purposes of leaving others unsaid.

The above-quoted passages all feature situations in which a character mismanages some major life event or trying set of circumstances. As Pascal notes, this is characteristic of free indirect style: "Situations of tension, of crisis, almost always rely on the reproduction through FIS of the character's view of the problem."²⁸ But free indirect discourse also works to mark the less fraught, more ordinary ways in which the imagined contours of an individual self are created, tended, policed, and propped. When the narrator of Henry James's *Washington Square* tells us that "Dr. Sloper would have liked to be proud of his daughter; but there was nothing to be proud of in poor Catherine," we understand that while Catherine may be the object of her father's criticism, the disdain voiced in *his* word, "poor," makes him the unwitting object of ours.²⁹ Indeed, "poor" is often a dead giveaway for free indirect discourse, especially in the nineteenth century, since it expresses a social rhetoric of sympathy that only thinly veils less acceptable feelings like contempt or disgust. The construction "would have liked to be proud of his daughter" conveys something distinctive about the doctor's inner life by indicating the self-camouflaging force of his need to make *his* lack of generosity *her* fault. Is it really true he would have liked to be proud of Catherine? Or doesn't he seem to rather sadistically enjoy his superiority over her? There is at least a suggestion, in other words, that this is the set of words Dr. Sloper employs not to deliver a candid opinion about his daughter but to construct a pose of generous resignation that lets him keep his own less admirable desires and motivations out of awareness. Dr. Sloper wishes to have the power and total knowledge of a third-person narrator; but because the actual narrator can secretly slip into and out of his telltale idiom, it reveals the stock formulations, concessions to vanity, and fixed ideas through which he constructs his false sense of a sovereign point of view. Here, as it so often does, free indirect discourse exposes pretension, delusion, and the self-generated myths through which reality is both managed and evaded.

In all of these examples, free indirect discourse functions as an instrument of critique and correction, a mode D. A. Miller sums up in his discussion of Jane Austen's *Emma*: "When free indirect style mimics Emma's thoughts and feelings, it simultaneously inflects them into keener observations of its own; for our benefit, if never for hers, it identifies, ridicules, corrects all the secret vanities and self-deceptions of which Emma, pleased as Punch, remains comically unconscious. And this is generally what being a character in Austen means: to be slapped silly by a narration whose constant battering, however satisfying—or terrifying—to readers, its recipient is kept from even noticing."³⁰ But it is also frequently used in what seem like very different kinds of situations: in moments of existential threat or emergency, as when Maggie Tulliver finds herself getting swept away by the river at the end of Eliot's *The Mill on the Floss*: "Great God! There were floating masses in it, that might dash against her boat as she passed, and cause her to perish too soon. What were those masses?"[31] The exclamation and the question mark are, clearly enough, expressions of Maggie's own sense of confusion and alarm; the passage thus seems to pull almost in the opposite direction of the kind of socially corrective function we see in the example from *Washington Square*. There, the narration is mocking and remote; in *The Mill on the Floss*, on the other hand, it is sympathetic and immediate. Yet both examples share what Pascal calls "the contrast between reality and the unavoidable and perhaps admirable self-delusions of persons imprisoned in the dimensions of individuality, time, and place."[32] That is, where the first version mines all the ironic humor inherent in a condition of self-imprisonment the character unknowingly expresses, the second takes us to the limit where the fictions of self are no longer enough to fend off the demands of reality. Both Dr. Sloper and Maggie Tulliver are trapped being when and where and what they are; the difference is that, for Maggie, the fatal implications of this condition can be denied no longer.

But of course, even this kind of palpable immediacy still keeps the character at something of a distance, not in an arch and ironic way, to be sure, but through the very expression of pity. For Freedgood, this distantiating effect is central to free indirect discourse and a key part of its ability to conjure reality:

> But at the moment of free indirect discourse, the narrator splits away from that creation; an act of disavowal allows a narrator to remove herself from her own creation, giving us the characters we think of as people—even though they are made up of textual bits and pieces that the narrator nets into sentences—the ones with whom we identify, however mistakenly, however correctly.[33]

Those textual bits and pieces throw into relief the various ways in which the contours of an individual self are imagined into being through language. This can also be true, of course, of first-person narration, especially of the "unreliable" variety. But what gives free indirect discourse its unique ability to render states of denial is the ubiquitous, hovering presence of the "omniscient" third-person narrator, which not only commandingly (if often only implicitly) maintains the claims of a wider social reality but also, in so doing, makes it feel as though those claims must be—or at least could be—apprehensible on some level by the individual character. The individual perspective, no matter how deluded or self-involved, always exists in the same "medium" as—even seems to be permeated by—something wider, more impersonal, and more capacious, beyond itself. Even Miller's asymmetrical vision that it is Emma's "self-deceptions" that the narrator mockingly corrects, quietly suggests that she is *on some level* aware of the trick she plays on herself. Because free indirect discourse always draws in indefinite measure from these different perspectives at once, because, as Casey Finch and Peter Bowen put it, "we can never know precisely who speaks in the free indirect style," the sense of "splitting away" Freedgood identifies never involves an entirely clean break along the ontological fault line dividing character and narrator; instead, the possibility always exists that the character is adopting an ironic attitude toward herself, perhaps even sees her own perspective from within and without at once.[34] As Ferguson argues, "Free indirect style is the stylistic equivalent of an amalgam of the individual speaking voice of a character and what that character would say if deducing a position from a larger perspective."[35] In some cases, the mixed register suggests a mixed state of mind, of different and undefinable levels of consciousness, forms of knowledge, self-awareness and self-obfuscation, operating simultaneously.

Consider, for example, this moment from *Jude the Obscure*, where Jude, at a church service, is inwardly chastising himself for the way he has let his sexual urges lead him off the path of righteousness and self-discipline:

> What a wicked worthless fellow he had been to give vent as he had done to an animal passion for a woman and allow it to lead to such disastrous consequences; then to think of putting an end to himself; then to go recklessly and get drunk. The great waves of pedal music tumbled round the choir, and nursed on the supernatural as he had been, it is not wonderful that he could hardly believe that the psalm was not specially set by some regardful Providence for this moment of his first entry into the solemn building. And yet it was the ordinary psalm for the twenty-fourth evening of the month.[36]

The litotic pileup of "it is not wonderful that he could hardly believe that the psalm was not specially set" perfectly captures the impacted equivocations of

Jude's state of mind. Unwilling to directly claim that his life is guided by a benevolent Providence—by this point in the story, he has plenty of evidence it isn't—he nevertheless allows himself to feel the tug of such a notion and live for a moment in its reassuring, half-spoken glow. "Nursed" is a particularly poignant word for this motherless child, one that suggests his long-standing condition of deprivation and his need for some source of consolation and nourishment. There is, in other words, a good deal of sympathy mixed into this critical portrait of the character's denialism. The mixed voicing of course ironizes Jude's point of view by allowing his characteristically hedged, half-in, half-out rhetoric to inflect the third-person narration. But it also introduces an ambiguity, even an instability, in the voice that brings out further complexities and dimensions to his state of mind. Because free indirect discourse combines, without distinctly marking off, the voice of the character and the voice of the narrator, it is often not entirely clear where one ends and the other begins. So "what a wicked worthless fellow" is clearly Jude's own self-abasing language, not the narrator's judgment of him. But what about that final, demystifying sentence: "And yet it was the ordinary psalm for the twenty-fourth evening of the month"? One might say this is the omniscient narrator reasserting his superior knowledge position (as he often does) and delivering the sobering truth that the character either cannot, or doesn't want to, see—that it is mere chance that has produced this convergence. Hardy *does* like dropping rhetorical hammers from on high like this. At the same time, however, it is also entirely possible that we can hear in that "and yet" traces of Jude's voice, reminding himself (as *he* often does) about the indifferent universe that repeatedly rebuffs his timid bids to distinguish himself. It's not at all clear, but that lack of clarity is not a flaw in the narrative voice but rather a vivid expressive effect—a means of making us feel the necessarily undefined border between inner and outer worlds and the hazy in-between zone that characterizes a state of denial.

The novel form is particularly good at representing such mixed, divided, or uncertain states of knowing and feeling like denial, because it can show the mind in continual, ongoing, dynamic negotiation with what George Eliot calls "the hard unaccommodating Actual"—the world as it is scraping against the world as the character wants it to be.[37] In Jude's case, we feel the clash between the character's felt experience of his own centrality and agency and an understanding of reality, grounded in Darwinian thinking, in which the individual life counts for little and is shoved here and there by biological drives that defy conscious control. We recall that the whole reason Jude has come to the church is to catch a glimpse of (we might say "stalk") his cousin Sue, which scheme, he keeps trying to convince himself, is motivated by something higher than mere "animal passion." That is, the entire situation is structured by denialism, and the church service functions as both a scheduled opportunity to track an object of desire and a zone in which motivations of this sort are conventionally disguised. Hardy's sympathy for Jude's denialism is underwritten by his deep loathing of the larger culture of denial that readily supplies confused human beings with artifices—consoling narrative patterns, stock phrases, swelling hymns—that insinuate themselves into their minds and allow them to evade the truth of their own experience. Stanley Cohen writes that "denial is always partial; some information is always registered. This paradox or doubleness—knowing and not knowing—is the heart of the concept."[38] This "doubleness" suggests an additional layer to the concept of a dual voice—it is not simply the combination of two distinct voices into one grammatical unit but also the production of a new kind of ambiguity that makes the very idea of distinct voices and knowledge positions immensely more complicated than had first appeared.

ELIOT'S HUMAN VOICES

A list of characters who dwell in some state of denial in *Middlemarch* would almost just be a list of all characters in *Middlemarch*. Casaubon, for example, *knows* his life's work, "The Key to All Mythologies," represents an intellectual dead end because its fatal flaws are simply there to be seen. Even a dilettante like his cousin Ladislaw spots them, including the uncomfortable fact that Casaubon does not read German, the language in which the most significant work in his field is being done. The work it takes to maintain this state of denial comes to define Casaubon's marriage to Dorothea, whose innocent queries aggravate him to the extent to which they threaten to expose what he already secretly knows. The banker Bulstrode, as another example, understands that his position of respectability and wealth in the town arises from his deceitful past practices; to preserve his own public standing as well as his own pious self-image, he must deny that unseemly history when it reappears in the person of his former factotum Raffles. "I shall decline to know you," he warns the man—meaning, of course, that he will refuse to acknowledge publicly their former association but also indicating the paradoxical force of his own long-standing denialism, the idea he can choose not to know that which he already knows.[39] Of all the characters in the novel, Bulstrode is the one whose denialism is most the strenuous because it has the biggest job to do: obscure the stark contradiction between his self-construction as a humble man of God and the actual truth of his fraudulence, pride, and violence. Not coincidentally, his voice, of all those in the novel, is the one most frequently rendered by way of free indirect discourse. More on this in a moment.

But it is the depiction of Dorothea Brooke that we find the novel's most complex and multidimensional representation of denial, because Dorothea's heroic idealism makes her susceptible to wishful thinking, while her acute reflexivity and penchant for sophisticated, abstract thought makes it difficult for her to dwell for long in such a state. In many ways, she is the inverse of Bulstrode: she doesn't bend reality to fit a presupposed righteousness but wants to test herself against reality in order to improve herself. We have already seen how tricky it can be to draw sharp divisions between a character's idiom and that of the narrator, and Dorothea's intense self-awareness makes this even more difficult. She is always, on some level, thinking of her own life as something that is being simultaneously lived and plotted, as if impersonally, within a larger imagined whole. Her voice, free indirectly rendered, seems to bend toward the narrator's wider perspective even, or especially, in those moments when she reveals her own blind spots:

> She felt sure that she would have accepted the judicious Hooker, if she had been born in time to save him from that wretched mistake he made in matrimony; or John Milton when his blindness had come on; or any of the other great men whose odd habits it would have been glorious piety to endure; but an amiable handsome baronet, who said "Exactly" to her remarks even when she expressed uncertainty,—how could he affect her as a lover? The really delightful marriage must be that where your husband was a sort of father, and could teach you even Hebrew, if you wished it.[40]

As Derek Oldfield says of this passage, "Part of the tragedy of Dorothea in these early chapters is conveyed by this quality of her thinking that we feel in its accuracy to come so near uncovering its own wrongheadedness."[41] Or, as the *Middlemarch* narrator has it a bit later in the novel, Dorothea keeps up a "sense of busy ineffectiveness, as of a dream which the dreamer begins to suspect."[42] With apologies to the Beatles: although Dorothea feels as if she's in a play, she is anyway. How judicious could Hooker have really been, Oldfield wonders, if he made such a "wretched," life-defining, mistake? The epithet "judicious" functions almost fetishistically, a marker of the discourse of thoughtful opinion that both conceals and exposes the absence of actual thought. The contradiction is there to be seen, riding along the surface of the language, and we sense that Dorothea, on some level, *does* see it. In this passage, as elsewhere in *Middlemarch*, Eliot uses free indirect discourse to pose rhetorical questions that simultaneously function as real ones:

> How could he affect her as a lover?[43]
>
> At this moment she felt angry with the perverse Sir James. Why did he not pay attention to Celia, and leave her to listen to Mr. Casaubon?[44]
>
> Before he left the next day, it had been decided that the marriage should take place within six weeks. Why not? Mr. Casaubon's house was ready.[45]
>
> The *fad* of drawing plans! What was life worth—what great faith was possible when the whole effect of one's actions could be withered up into such parched rubbish as that?[46]
>
> How could it occur to her to examine the letter, to look at it critically as a profession of love?[47]

Such questions are ostensibly declarations of a firm position, but they also *actually* ask things that lead into uncomfortable territory. How could the hale and "red-faced" Sir James affect her as a lover? The answer is obvious: sexually. (Which is also why he is paying attention to her instead of Celia!) What is life worth if one's actions have little effect on the world or if one's loftiest motivations never come uncontaminated with some amount of self-interest? Dorothea here half poses and half dodges the idea that belief may be nothing more than the need to believe, the expression of a desire to see life as meaningful out of fear of facing the emptiness of the alternative. How could it occur to her to read Casaubon's letter critically? The fact that she's asking the question means it just did! In that last example, we see perhaps most clearly the subtle purposes to which the ambiguity of free indirect discourse can be put: the fact that we cannot quite tell whether this is the narrator making excuses for her character's startling lack of skepticism, or if it's the character half-reflecting on her own motivations, produces a chiaroscuro-like effect, the fade of ambiguity adding a sense of depth and dimensionality to the depiction of consciousness.

Pascal identifies this effect as a crucial feature of Eliot's late style: "Dorothea Brooke or Daniel Deronda . . . enjoy a high degree of self-awareness and think and speak in terms very close to that of the narrator."[48] The result, as he points out, is not just the kind of blurring effect that always attends free indirect discourse but an additional layer of uncertainty whereby we are often unable to recognize where free indirect discourse *itself* begins and ends. Along with her cerebral characters who think and speak like narrators, Pascal notes, her narrator has a tendency to express emotions, voice questions, and make exclamations in ways that sound character-like: "Uncertainty of discrimination between the narratorial and the subjective perspective arises from the frequent use, for authorial comments, of the signals, the indices, of free indirect speech"; such "intrusions," he notes, become especially difficult to demarcate in any definitive way "when they occur in the neighbourhood of true FIS passages."[49]

The tension thus generated by free indirect discourse, the sense that Dorothea always seems at least potentially aware of her own self-generated illusions and can "hear" the unresolved questions or problematic implications of her own positions, is itself put in motion over the course of the eight hundred or so pages. We sense that she does not quite know, then refuses to know, then pretends not to know, then finally does know, that her husband is not the man she imagined him to be and that, actually, the entire world of men is kind of a giant sham. In fact, part of the great drama of *Middlemarch* is watching her pass through these states of mind and relationships to knowledge as the reality of her situation with Casaubon and everything it implies becomes increasingly difficult to deny. It also helps produce the sense of telos in the novel: the idea that a condition of denial is gradually (though imperfectly) dispelled, and she comes to recognize what, on some level, she always knew. In Dorothea's case, this drama is compounded by the fact that there is seemingly no end to the terrible things to know about Casaubon: as soon as one defect is acknowledged, something else, something even more disturbing and fundamental, presents itself. Dorothea first has to accept that the "Key to All Mythologies" is an intellectual dead-end (and thus she has not married Milton or Hooker); then she has to accept that his position with regard to his cousin Ladislaw is defined by prejudice, enmity, and, in a sense, theft (and thus she has not even married a particularly good man); *then* she has to accept that he was willing to smear her good name and constrain her choices for life (and thus she has married a *terrible* man). There is a sense of inevitability and momentum to the downward progression, even as there is a countervailing desire to draw a line and declare the end reached. But while the long descent into the cellar of Casaubon's soul has some clearly plotted steps where genuinely unknown information makes certain realities newly or differently visible (the surprising disclosure of the terms of his will, for example), each step is also anticipated via the blending of knowledge positions communicated through the expressive ambiguity of free indirect discourse. The result is that revelations in the novel almost always feel like they come from within and without at the same time.

The manner in which free indirect discourse highlights the fetishistic nature of language—how it conceals by seeming to reveal, vocalizes for the purposes of muffling, advertises its failure as a defense mechanism as it gets repeatedly conscripted for just that purpose—is key to Eliot's ethical project in *Middlemarch*. One of her main targets is the way in which reality gets misnamed through the use of conventional discursive structures and terms; such misnaming paves the way for misunderstandings born out of admirable intentions or naïveté (e.g., "judicious") but also makes possible all manner of bad-faith behavior and allows self-interest to gather beneath the banners of piety. This is clearly the case with Casaubon, to whom the word "generous" and its variants affixes itself in the early chapters of the novel, despite behavior that is anything but. Or with Bulstrode, whose tortured self-justifications are consistently rendered through long passages of free indirect discourse: "It was an hour of anguish for him very different from the hours in which his struggle had been securely private, and which had ended with a sense that his secret misdeeds were pardoned and his services accepted. Those misdeeds, even when committed—had they not been half sanctified by the singleness of his desire to devote himself and all he possessed to the furtherance of the divine scheme? And was he after all to become a mere stone of stumbling and a rock of offence? For who would understand the work within him?"[50] Free indirect discourse not only makes evident his own internal divisions; it also offers itself as silent judge and eavesdropper on a private rhetoric of self-deception, the putative "singleness" of his desire cleverly and continually belied by the doubleness of the voice expressing it.

The obfuscation created by such appeals to the "divine scheme" roughly fits with what Stanley Cohen calls "interpretive denial"—not so much the refusal to acknowledge a given reality but the tendency to euphemize or otherwise rename it until it squares with one's preferred version of the world.[51] For Eliot, religion *is* interpretive denial, not simply because it provides the terms and concepts through which life can seem meaningful but also because it allows one to seize a kind of fictitious private dominion over language so as to render its shared, public power and value entirely a matter of internal self-dealing. She says of Bulstrode: "It is only what we are vividly conscious of that we can vividly imagine to be seen by Omniscience," suggesting, in full Feuerbachian spirit, that there is no distinction between personal and divine moral ledgers because there is no distinction between self and God.[52] For Eliot, as J. Hillis Miller argues, "God does not exist except as an individual or collective projection. It is therefore possible, in a false imaginary transaction with this nonexistent deity, to rename any bad act a good one by a species of metaphor or metamorphosis, as stolen money is 'laundered' in another country."[53] Free indirect discourse, in a sense, lays bare the mechanism whereby characters pass off counterfeit bills to themselves because it embeds such closed transactions within an open system of meaning production. As we have seen, it puts character idiom and social discourse, private and public, into a state of necessary and irresolvable tension. I note in passing how critics sometimes use the metaphor of disputed "ownership" to talk about the uncertain status of language in free indirect discourse. Take, for example, James Wood's description in *How Fiction Works*: "The narrative seems to float away from the novelist and take on the properties of the character, who now seems to 'own' the words."[54] That very desire for "ownership" has, in Eliot's hands, a clear moral dimension, insofar as it communicates a character's need to *dis*own some unsettling but necessary truth about his or her actions, desires, or circumstances.

In a sense, Eliot's point is not simply that the imagination of a surveilling God creates the conditions for denialism to take root but that a state of denial is what creates and sustains the idea of a surveilling God in the first place. The narrator makes the circularity evident: "His equivocations with himself about the death of Raffles had sustained the conception of an Omniscience whom he prayed to."[55] The only thing that has the power to break this self-enclosure is the restoration of language to the social, secular sphere, where meaning is not subject to this kind of personal tyranny. The passage continues:

> Yet he had a terror upon him which would not let him expose them to judgment by a full confession to his wife: the acts which he had washed and diluted with inward argument and motive, and for which it seemed comparatively easy to win invisible pardon—what name would she call them by? That she should ever silently call his acts Murder was what he could not bear. He felt shrouded by her doubt: he got strength to face her from the sense that she could not yet feel warranted in pronouncing that worst condemnation on him. Some time, perhaps—when he was dying—he would tell her all: in the deep shadow of that time, when she held his hand in the gathering darkness, she might listen without recoiling from his touch. Perhaps: but concealment had been the habit of his life, and the impulse to confession had no power against the dread of a deeper humiliation.[56]

If Dorothea is an obscure St. Theresa, Bulstrode is a shoddy St. Augustine, and his own version of "make me pure—but not yet" will very clearly keep him trapped in his own bad habits. Those habits, like Augustine's vices, are woven deeply into his psyche, having everything to do with the paranoid maintenance of the sense of self he has taken such pains to construct. But we also see that what he truly fears here is not God's judgment—that can always be verbally managed—but a world in which reality gets properly named and where that naming is not in his control. For Hillis Miller, this is at the heart of Eliot's ethical project: "Correct judgment is defined as correct denomination, naming the person and his deeds correctly as what they are. To call a person or a deed by its right name is proper realism or faithfulness in narration."[57] I would add that "proper realism" inheres not just in correct naming practices but also in the identification of the motivated process by which the incorrect ones get lodged in consciousness.

Eliot dramatizes these power dynamics through her use of voice. The remark, "some time, perhaps—when he was dying—he would tell her all," is clearly Bulstrode's own wishful thinking, his denialism rendered, as usual, via free indirect discourse. But the final sentence of the passage—"Perhaps: but concealment had been the habit of his life"—is different. This sounds much more like the narrator (although, as always, it's hard to be entirely certain), who dispels such illusions and seems to know the character better than he knows himself. Bulstrode's "perhaps" introduces a kind of indefinite vagueness that serves his purposes of obfuscation and deferral; the narrator's "perhaps," in contrast, functions as an ironizing echo that sounds the corrective note of skepticism. The word thus gets reclaimed through repetition: wrested from the character's closed system of internal meaning making, it is newly wielded in a way that overrides his control. Such a reclamation suggests that although we cannot entirely foreclose the possibility of Bulstrode's eventual redemption, we also can't take his word for it.

Although *Middlemarch* is a work of high realism, it takes notable excursions into other generic modes, including, and especially, the gothic. Dorothea, for example, facing the prospect of finishing her husband's scholarly project after he dies, imagines that task as a kind of premature burial: "And now she pictured to herself the days, and months, and years which she must spend in sorting what might be called shattered mummies, and fragments of a tradition which was itself a mosaic wrought from crushed ruins—sorting them as food for a theory which was already withered in the birth like an elfin child."[58] And Raffles appears to Bulstrode like a rum-soaked Mephistopheles, returned to exact a toll for the sinister offices he undertook on his patron's behalf: "As if by some hideous magic, this loud red figure had risen before him in unmanageable solidity—an incorporate past which had not entered into his imagination of chastisements."[59] The originating "sin" committed by these two characters is the belief in the primacy of their own life stories within the larger scheme of reality, and in both cases, the corrective is felt through, and described in, a gothic idiom. Dorothea's version of this is of course much more modest than Bulstrode's, who has an "immense need of being something important and predominating," but her assumption that the local pedant is a Hooker or Milton testifies to a quietly egocentric belief that her life takes shape in intimate proximity to the world-historical.[60] But a paradox of sorts arises in the turn to the gothic to deliver this sort of comeuppance, which is that the punishment meted out for believing the world pays you special attention feels like an allegorically appropriate nemesis conjured up just for you. As Carroll puts it, "The mind in its pride seeks to redeem by fiat the fallen world in which it lives, but instead turns it into an inferno where it is hunted down by monsters of its own creating."[61] But of course, Bulstrode isn't Dr. Frankenstein, and Raffles wasn't his "creation": for him to see the latter as some avenging agent bodied forth to bring "chastisement" from a disapproving universe is merely to invert the terms of his providentialism and preserve the foundational epistemological error. As Carroll points out, the problem for Bulstrode is that Raffles, like Frankenstein's creature, has an inconvenient agency of his own; but any expression of that agency immediately gets converted into a narrative of persecution and punishment that maintains Bulstrode as the cynosure. He would deny Raffles an independent existence, but the very methods by which Raffles would contrarily express that independence perversely strengthens, rather than vanquishes, his state of denial.

Dorothea suffers from a milder case of this kind of thinking—the gothic idiom is employed less vividly and with considerably less paranoia—and her way out of it occurs via a reckoning with her own existential obscurity and a concomitant acceptance of others as equivalent centers of self; that is, it occurs via an embrace of what Eliot codes as "realism." This is the "new consciousness of interdependence" Dorothea spends the novel working to achieve: a perspective oriented horizontally, toward multitudinous, dynamically related coexistent perspectives rather than vertically and hierarchically, toward divine projections and individually centered allegorical dramas.[62] It is thus rooted in an emergent secular, scientifically minded worldview (hence the "new") in which "personality is an obstruction to perception," as George Levine puts it, and the desiring self must be recognized as an engine of distortion and denial.[63] In her letters, Eliot's way of describing this has a distant ecological ring:

> I try to delight in the sunshine that will be when I shall never see it any more. And I think it is possible for this sort of impersonal life to attain great

intensity,—possible for us to gain much more independence, than is usually believed, of the small bundle of facts that make our own personality.[64]

For Eliot, as exemplified in many of her protagonists, the ethical life depends on moving from the "moral stupidity" that views "the world as an udder to feed our supreme selves" to a concern for others and the larger social totality in which self-interest is sidelined to whatever extent possible.[65] As she tends to do, Eliot fuses epistemology and ethics, positing that a correct moral orientation toward the world produces a clearer vision of it, and vice versa (as Levine puts it, "in a world so imagined, intelligence and morality are interdependent").[66] This suggests an additional layer of significance in her use of gothic tropes, insofar as horror inheres not in getting singled out by devils and poltergeists but in the way feeling so haunted speaks to a condition of enclosure within selfhood—the choices one has made in the service of self as well as the bottomlessness of the need to continue serving it. At the end of the novel, Dorothea famously overcomes her own intense jealousy to aid the woman she mistakenly believes to be her romantic rival:

> She was beginning to fear that she should not be able to suppress herself enough to the end of this meeting, and while her hand was still resting on Rosamond's lap, though the hand underneath it was withdrawn, she was struggling against her own rising sobs. She tried to master herself with the thought that this might be a turning point in three lives—not in her own; no, there the irrevocable had happened, but—in those three lives which were touching hers with the solemn neighbourhood of danger and distress.[67]

Note that Dorothea's act of self-denial touches off Rosamond's awakening out of her own state of denial: "She was under the first great shock that had shattered her dream-world in which she had been easily confident of herself and critical of others. . . . [It] made her soul totter all the more with a sense that she had been walking in an unknown world which had just broken in upon her."[68] Such a moment speaks to the horizontal and secular moral orientation of the novel, in which selfless acts can have outwardly ramifying, undesigned consequences that are entirely explicable according to the logic of causality, verisimilitude, and ordinary human psychology. And note, finally, how Dorothea's decision to commit this selfless act is communicated via free indirect discourse, which occurs after the interruptive dash: "not in her own; no, there the irrevocable had happened." The irrevocable has actually *not* happened, and thus we are in the familiar territory where free indirect discourse affords a glimpse of a mistaken perspective precipitated out of powerful emotion. But this also signals that a key shift has occurred: reality-denial has been converted into a form of self-denial, Dorothea erring on the side of resignation rather than wishful thinking, the surrender of self rather than its amplification. Her movement toward what Eliot describes as the "impersonal life" means a movement toward the perspective of the narrator, what Catherine Gallagher calls the "generic imperative that the protagonist's being should come to resemble the uncharacterizable universal consciousness of the narrator and implied reader as the novel progresses."[69] The gap between character and narrator closes over the course of the novel, the diminishing distance pinged periodically by the sonar of free indirect discourse. In this way, we can identify the progressive telos embedded in and expressed by the dynamic—we might even say "evolving"—relationship between dual narrative voices over the course of the novel.

And yet it's important to also note that despite the imperative to divest oneself of ego-generated illusions, *Middlemarch* also contains a countervailing impulse to preserve certain forms of unknowing. What can and perhaps *must* be denied a place in the forefront of consciousness, she suggests, is the non-human world. In fact, "suggests" is the wrong term—she just says it in one of the novel's most famous passages: "If we had a keen vision and feeling of all ordinary human life, it would be like hearing the grass grow and the squirrel's heart beat, and we should die of that roar which lives on the other side of silence. As it is, the quickest of us walk around well wadded with stupidity."[70] George Levine writes that, for Eliot, making an effort to tune in to the other side of silence is an ethical imperative, in part because attending to the "absolute otherness of things" allows us to push past the narrow bounds of self and its moral stupidity.[71] And yet this passage also seems to make the countervailing suggestion that humans are unequipped psychologically, perhaps even physiologically, to handle such an experience of otherness, so that being "well-wadded with stupidity" is, on some level, not only a normal condition but a necessary one. As Ivan Kreilkamp writes, Eliot here sets a "metaphorical outer limit" to sympathetic attention, one posited along the imagined boundary between human and non-human.[72]

Or, we might say, posited along the boundary where the human *becomes* the non-human, at the place where human life, imagined as an "all," an undifferentiated aggregate, loses its distinctly human quality and becomes mere white noise as raw and elemental as any other blind organic phenomenon. The reason not to meditate on such a world is practical and self-protective: to tune into it is to be overwhelmed to the point of death, and, anyway, we are not meant to hear it because the human sensorium lacks the capacity to handle that kind of input (signaled through the illogical mix of sensory registers, of a "keen vision" perceiving an auditory phenomenon). If the voice of free indirect discourse functions as the differentiating sounding board of social correction and moral development, this roar is simply raw vocality: inhuman and indiscriminate. Impossible to handle, impossible to perceive, except perhaps through the momentarily alienating jolt of a thought experiment. Here, knowledge is decoupled from its usual place as morality's helpmate: where elsewhere we had the problem of the "moral stupidity" of personal egotism, here we are told that being "well-wadded with stupidity" is actually a necessary condition for preserving sanity, intelligibility, and the contours of the human. Here, in other words, is the case *for* denial: a confession of realism's limits but also a justification of where they should be drawn.

A similar transformation of the human aggregate into something non-human occurs in *The Mill on the Floss*, where the narrator muses on the "oppressive feeling" caused by the sight of villages along the Rhône, which appear to her like a "gross sum of obscure vitality that will be swept into the same oblivion with the generations of ants and beavers."[73] In this case, Eliot doesn't so much close down the subject as change it. As she shifts scenes from one river to another, from Rhône to Floss, she also quietly shifts the terms of the debate: "Perhaps something akin to this oppressive feeling may have weighed upon you in watching this old-fashioned family life on the banks of the Floss.... It is a sordid life, you say, this of the Tullivers and the Dodsons, irradiated by no sublime principles, no romantic visions, no active, self-renouncing faith."[74] What was initially oppressive about the Rhône was the vision of purposeless organic fecundity and waste, but that is very different from what is wrong with life on the Floss: an unromantic and undistinguished existence. "Obscure vitality"—the ruptured boundaries of the sphere of the human—is decidedly not the same as social obscurity, an unnoticed existence *within* that sphere. The word "sordid" carries us back inside the borders, effecting a shift from the merely material back to the more manageable zone of human value judgments and distinctions. And since neither family seems to warrant quite such a harsh epithet, the reader ("you") is encouraged to resist the interpellation, to insist upon the kinds of careful, smaller-scale moral and social distinctions that, just a moment ago, were in danger of getting washed away.

Both passages—but especially the one from *Middlemarch*—are unusually self-aware examples of what the environmental critic Val Plumwood calls the "backgrounding" of the non-human world, which emphasizes the "denial of dependence on biospheric processes, and a view of humans as apart, outside of nature, which is treated as a limitless provider without needs of its own."[75] Of course, Eliot's chosen visions of the non-human background seem far removed from questions of "human dependence on biospheric processes"—a soundscape of rodent hearts and groaning grass fibers is so freakish that it's hard to think of it as actually representing "nature" in any meaningful sense. But in a way, that's the whole point. When we momentarily tune in to a natural world, it does not seem to be something on which anything depends, but something more like a kind of separate, even parallel, reality pulsing along to its own bizarre rhythms and noises.

The "roar on the other side of silence" seems to have been inspired by a section of T. H. Huxley's *On The Physical Basis of Life*, in which he gives this description: "The wonderful noonday silence of a tropical forest is, after all, due only to the dullness of our hearing; and could our ears catch the murmur of these tiny Maelstroms, as they whirl in the innumerable myriads of living cells which constitute each tree, we should be stunned, as with the roar of a great city."[76] We can see in *Middlemarch* how Darwinian science and a "new consciousness of interdependence" can be used in the service of an emergent secular humanism that exposes the self-dealing denialism of Pharisaical believers like Bulstrode or even the less harmful desire for "glorious piety" that initially motivates Dorothea. But Darwinian naturalism and its vision of "the physical basis of life" also carry with it obvious implications that are profoundly antithetical to the entire humanist project. That threaten, indeed, to drown it out. The wall of noise and the rushing tides of generations: these too are visions of "impersonality" that actually follow rather naturally from the horizontal, decentering scientific perspective Eliot elsewhere embraces. One obvious conclusion to be drawn from developments in evolutionary biology is that the individual organism exists simply as a vector of drive, motion, and urge in a larger system of procreative compulsions and protein transfers. If we pose difficult questions about a given individual's sense of exceptionalism, what stops us from asking such questions on the level of the species as well? Lurking behind both passages—though more explicitly on offer in *The Mill on the Floss*—is the problem of extinction, the troubling notion that the universe is not in any way ordered around, in correspondence with, or dependent upon the human. Our purchase on it, in both temporal and perceptual terms, is partial, slight, and fleeting; for that reason, the non-human can be both acknowledged as "real" and strategically pushed out of mind as not "realistic." As we've seen, this is the question she explicitly poses to herself in "Leaves from a Note-Book": "Is it not conceivable that some facts as to the tendency of things affecting the final destination of the race might be more hurtful when they had entered into the human consciousness than they would have been if they had remained purely external in their activity?"[77] The problem of extinction, which any reckoning with the full ecological embeddedness of *Homo sapiens* in the natural world must confront, looms not simply as an object of dread but as something harmful, perhaps better kept out of sight. Or out of earshot.

If, as Kucich argues, Eliot was "appalled by man's universal and irreversible amalgamation with matter," both passages deploy the procedures of realism to restore a reassuring sense of separation. The horror of the gothic and its assorted vampires, demons, and succubi can be readily dispatched by the ironic, remediating procedures of narrative realism; here, though, is horror of a different sort, one that is more difficult to contend with because it emerges from the same epistemological ground upon which realism itself is founded. Note the difference in narrative voice used to handle these different vectors of denial: Where free indirect discourse was elsewhere employed correctively and teleologically to mark and, in the case of Dorothea, to narrate the process by which illusions get gradually dispelled through the applied pressure of a consensus social reality, non-human scales and threats are here handled via a direct narratorial address to the reader ("it is a sordid life, you say"). In this case, the question about where to draw the line to refortify the boundaries of the human itself cannot be handled by free indirect discourse, which polices a boundary internal to the human experience. As John MacNeill Miller writes, such moments "enter the text only at the extradiegetic level as useful illustrations of the accepted, unproblematic exclusions already implicit in the social novel—exclusions that can now be reintroduced, symbolically, to remind the reader of the necessity of exclusion itself."[78] These meta-narrative moments make the case for the novel's own distribution of attention while admitting the necessary suppression of other possibilities. Such scenes are not part of the characters' moral education, nor, it is implied, do they need to be.

We might compare Eliot's approach to that of Darwin's in *The Descent of Man*, which appeared during the serial run of *Middlemarch*. The advance of moral evolution, Darwin argues, is to be found precisely in the expansion of the boundaries of sympathy beyond the species barrier: "As man gradually advanced in intellectual power, and was enabled to trace the more remote consequences of his actions; as he acquired sufficient knowledge to reject baneful customs and superstitions; as he regarded more and more, not only the welfare, but the happiness of his fellow-men; as from habit, following on beneficial experience, instruction and example, his sympathies became more tender and more widely diffused . . . and finally to the lower animals."[79] Just a few paragraphs earlier, he uses the same language to discuss "our sympathies becoming more tender and more widely diffused, until they are extended to all sentient beings."[80] As Ian Duncan succinctly puts it, "Darwin affirms the totality of (sentient) life as an ethical telos."[81] And yet, characteristically for this book, the radical extension of feeling is made possible, in part, by a refortification of the grounds of white European superiority. By showing themselves capable of acknowledging kinship across cultures, distances, and, ultimately, over the species divide, Europeans both prove themselves to be more highly evolved than their non-human brethren and distance themselves from "savages," for whom this "sympathy beyond the confines of man" goes "unfelt."[82] In other words, and as we have seen him do elsewhere, Darwin's radical leveling is put in the service of reestablishing a racial hierarchy. In this way, the passage serves as a perfect example of what Zakiyyah Iman Jackson, following Saidiya Hartman, describes as the racist logic that often structured sentimental arguments about extending sympathy to non-human animals: "The 'humane' is an ideal that suggests humanity is gained by performing acts of kindness.... Rather than forestall domination, 'humane' discourse, in effect, made human identity contingent on hierarchical relationality—encounters between those with refined sensibilities and those presumably without."[83]

Eliot avoids such rhetoric and the hierarchizing logic that structures it. For her, the "diffusion" of sympathy occurs through an embrace of the human world and its limited horizons, which enables an investment in the local and immediate. But preserving the reality of a moral world in a universe without a moral structure requires some suspension of disbelief, some amount of denial. J. Hillis Miller puts it this way:

> Ethical good for George Eliot is based not on sound judicial verdict but on a performative fiat. It is an act of arbitrary renaming which has no more solid base than the religious denominations it replaces. It therefore is precariously balanced in the air, vulnerable to the knowledge which would make it impossible. It is always endangered by just that insight her dismantling of religious naming gives.... It simultaneously shows the reader he can and must know, and shows him he cannot and must not know. The "can" and "must" here are both an ethical and a linguistic imperative. We should know and not know, and we cannot help both knowing and not knowing.[84]

Believing in one's "duty" despite the ultimate meaninglessness of the term; behaving generously and sympathetically toward others by not thinking about all the ways in which they might not deserve it; pursuing the "good" irrespective of its efficacy, whether it is noticed, whether it really matters: such acts of "both knowing and not knowing" compose a kind of denialism that is a familiar, widely practiced, perhaps even necessary part of social life. As John Miller puts it, "The sublime extent of interconnection could also generate a sense of insuperable ethical ambiguity: faced with the astonishing expanse and unknowability of interrelatedness, it is difficult to ascertain all the consequences of any individual action and decide its ethical import—making it easy to justify a quietist resignation, a refusal to do anything at all."[85]

But (as Miller also argues) even if we would agree that some amount of denial is required to construct or inhabit a coherent moral reality, what seems much less clear is whether the non-human world ought to be used as the *reductio ad absurdum* in an argument about the appropriate bounds of sympathy and attention. That is to say, if Darwin sidesteps the problem of human exceptionalism by doubling down on other, more destructive discourses of superiority, Eliot sidesteps it by pushing it beyond the representational boundaries of realism. As Helen Kingstone notes, "She ends up suggesting that despite Darwin, we have to maintain lines of demarcation. Despite the clamoring voices of other organisms, we can only function as human beings—and sympathize with our fellow humans—if we block them out."[86] This, then, is a form of the "soft" denial of Darwin: not an outright refusal to accept the scientific evidence, not a reassertion of impassable ontological boundaries between human and non-human, but rather a strategic decision to believe in "lines of demarcation" despite their evident fictitiousness. But this means that isolated moments in Eliot's fiction that might suggest the glimmerings of large-scale environmental concern—the idea of "the world as an udder to feed our supreme selves" in *Middlemarch*, for example—are, in context, mostly invoking the planetary for the purposes of hyperbole to highlight the monstrously outsized nature of the individual ego.[87] That is to say, the focused critique of individual exceptionalism both suggests and sidelines a more ecologically minded critique of species exceptionalism, of questions about the domination of nature and its wholesale conversion into resources.

What ultimately maintains these boundaries—and what, not coincidentally, governs the functioning of free indirect discourse—is history. For Eliot, the necessary fictions that make moral life possible are also at the root of her historicism; in fact, the moral life and the historical imagination are, for her, mutually constitutive, as the closing paragraphs of *Middlemarch* indicate: "Her full nature, like that river of which Cyrus broke the strength, spent itself in channels which had no great name on the earth. But the effect of her being on those around her was incalculably diffusive: for the growing good of the world is partly dependent on unhistoric acts; and that things are not so ill with you and me as they might have been, is half owing to the number who lived faithfully a hidden life, and rest in unvisited tombs."[88] In contrast to Darwin's sense of an increasingly "diffused" sympathy defined as a deliberate act of widening biophilic extension, the moral diffusion Dorothea occasions does not depend on her knowing that it happens, but it does depend on the in-built retentive capacity of the wider world. But "world" here most assuredly does not mean "planet," and "unhistoric" simply means "not chronicled." What goes unnamed and unrecognized still *matters*; the moral life radiates outward and is preserved in the movement of human history, quietly creating the future social and cultural "world" in which succeeding generations will find expanded possibilities for flourishing.

History imagined in this way, as the ultimate ground of meaning and value, depends upon what Dipesh Chakrabarty calls "the age-old humanist distinction between natural history and human history."[89] The flowing river at the end of *Middlemarch* is decidedly not the rushing water that sweeps away the generations of ants and beavers in *The Mill on the Floss*; it preserves and channels rather than obliterates. The reference to the reform-minded Cyrus the Great and his strategic engineering of the Euphrates River—an event that led to his relatively peaceful conquest of Babylon—suggests the beneficent power of human technological interventionism as a directive agent in the course of history: subduing and managing the rushing waters rather than getting swept away by them. Meanwhile, it links her own project to that of the main historian of that event, Herodotus, whose digressive, multivalent historiography also emphasized surprising interdependencies and located sources of historical significance in everyday experience. Such a vision represents not an "amalgamation" of the human with matter but a metaphor that frames human history as a giant, lossless energy system—a far cry from the actual regime upon which such notions of progress were being erected.[90]

The bifurcation of natural and human history is, as Chakrabarty says, an "age-old" distinction, appearing in such diverse historical thinkers as Vico, Hobbes, and Hegel. But there was new pressure on it in a period in which evolutionary science made any firm ontological division between the two impossible. Darwin and his fellow scientific naturalists reimagined the history of the human species as fully embedded in the same processes and pressures that shaped, for lack of a better term, the "natural" world: indeed, when, at the end of *The Origin of Species*, Darwin promised that in future work on evolution, "light will be thrown on the origin of man and his history," that last word makes biological and cultural development as part of one continuous process.[91] At the same time, however, and paradoxically, biological history and cultural history were widely imagined as functionally distinct insofar as the latter had enabled humans to gain control and subdue nature to such an extent as to make itself a nearly autonomous sphere. This distinction could be defined internally, in the ability of (some) humans to suppress their biological instincts, and externally, in the ability of "advanced" cultures to direct and shape natural processes to their ends. In his 1864 essay, "The Origin of Human Races," Alfred Russel Wallace writes:

> From the moment when the first skin was used as a covering, when the first rude spear was formed to assist in the chase, the first seed sown or shoot planted, a grand revolution was effected in nature, a revolution which in all the previous ages of the earth's history had had no parallel, for a being had arisen who was no longer necessarily subject to change with the changing universe—a being who was in some degree superior to nature, inasmuch, as he knew how to control and regulate her action, and could keep himself in harmony with her, not by a change in body, but by an advance of mind. Here, then, we see the true grandeur and dignity of man.[92]

In this passage (and in many others like it) language is used to inflect a difference in degree back into a difference in kind—there is a hint of the supernatural in the phrase "a being had arisen" and something of the biblical promise of human exclusivity and dominion in the phrase "superior to nature." And yet note that it is not just the ability of *Homo sapiens* to exert its will upon "nature" that distinguishes it from other animals but its ability to reflect on its ability to do this. The very ability to imagine a separate thing called "human history" is what makes it a separate thing, distinguishing not only humans from non-human animals but "civilized" from "savage." Mark Poster sums up the widespread assumption that "history and humanity are coterminous,"[93] quoting, as a key example, a passage from E. P. Thompson (who was himself referring to a famous line from Marx's *Eighteenth Brumaire*): "Men make their own history: they are part agents, part victims: it is precisely the element of agency which distinguishes them from the beasts, which is the *human* part of man."[94]

Eliot's vision of history rests on similar grounds. As Kingstone puts it, "For Eliot . . . our historicism is what makes us human."[95] This can be perhaps most clearly seen in *The Impressions of Theophrastus Such*, where she writes, "The divine gift of a memory . . . inspires the moments with a past, a present, and a future, and gives the sense of corporate existence that raises man above the otherwise more respectable and innocent brute."[96] That passage occurs in the last chapter in the book, "The Modern Hep! Hep! Hep!," which connects these questions of history, memory, and the definition of the human to race and empire. I now turn my attention to Eliot's last finished novel, *Daniel Deronda*, which offers an alternative vision of history and thus a different solution to the problems of denial and species exceptionalism posed by Darwin.

RACE, EMPIRE, AND *DANIEL DERONDA*

Early in the novel, Eliot poses a loaded question about the relationship between her protagonist Gwendolen and the larger historical forces sweeping through the world:

> Could there be a slenderer, more insignificant thread in human history than this consciousness of a girl, busy with her small inferences of the way in which she could make her life pleasant?—in a time, too, when ideas were with fresh vigour making armies of themselves, and the universal kinship was declaring itself fiercely: when women on the other side of the world would not mourn for the husbands and sons who died bravely in a common cause, and men stinted of bread on our side of the world heard of that willing loss and were patient: a time when the soul of man was waking to pulses which had for centuries been beating in him unfelt, until their full sum made a new life of terror or of joy.
>
> What in the midst of that mighty drama are girls and their blind visions? They are the Yea or Nay of that good for which men are enduring and fighting. In these delicate vessels is borne onward through the ages the treasure of human affections.[97]

Here is a key moment marking an important difference between *Middlemarch* and *Daniel Deronda*. In the latter novel, the epic movement of human history, here figured in the unfolding of the American Civil War and the battle over slavery, is set in deliberately awkward contrast with the small-minded story of one of the protagonists. The passage arises seemingly out of nowhere at the end of a chapter, apropos of nothing occurring at that moment in the diegetic world of the narration. Of course, this is the point:

the narrator has to clear her throat and ask such questions if they are to be raised at all, because they are utterly beyond the narrow concerns of the characters. The narrator's attempt, in the last paragraph, to reconcile the inconsequentiality of her subject with the motions of human history is obviously riddled with cant and sentimentality. We know full well that Gwendolen is anything but a simple container for "affections"—she's in fact something of a misanthrope—and that the Yea or Nay question works itself out in her plot not in grand Carlylean terms but as the restrictive binary of mate selection that she mistakes for actual power.

Where Dorothea is consistently imagined as a source of potential energy, Gwendolen appears for most of the novel as an unwitting agent of entropy and loss, both participant and victim of a social world defined by exhaustion. Early in the novel, Eliot remarks about her that, "a soul burning with a sense of what the universe is not, and ready to take all existence as fuel, is nevertheless held captive by the ordinary wire-work of social forms and does nothing particular."[98] Where Gwendolen's extractivist, world-devouring ego is checked by the many restrictions attending her gender, no such wire-work seems to impede Grandcourt, who is consistently described as a kind of bottomless-energy drain, using up money, resources, commodities, and other lives for the sake of whatever perverse sensations his ego demands. No accident that the discarded Lydia Glasher lives in a "country, once entirely rural and lovely, now black with coal-mines" covered with "black roads and black mounds which seemed to put the district in mourning."[99] This conjunction of discarded character with plundered environment is a striking illustration of Elizabeth Carolyn Miller's argument about the nineteenth-century provincial realist novel and the anxieties it expresses about "reproductive futurity" through the representation of domestic sacrifice zones.[100] The landscape is striking too in its dramatic departure from the usual settings and places featured in *Daniel Deronda*, an incongruity that makes its allegorical function impossible to miss. Later, we get shades of Governor Eyre and a glimpse ahead to Conrad's Kurtz in a brief counterfactual: "If this white-handed man with the perpendicular profile had been sent to govern a difficult colony, he might have won reputation among his contemporaries. He had certainly ability, would have understood that it was safer to exterminate than to cajole superseded proprietors, and would not have flinched from making things safe in that way."[101] Eyre makes the reference topical and, thus, not altogether surprising; still, there is something jarring about the sudden change in the frame of reference—the brief imagining of an entirely different life for Grandcourt and the splitting apart of the euphemism of "making things safe" with a vision of extermination. It's not the "wire-work" of convention that keeps Grandcourt from living out this destiny; it's simply an accident of history that he wasn't able to vent his sadism on the world in officially autocratic form.

Grandcourt is paradoxically positioned as both archetype and anomaly. He is, on the one hand, the very embodiment of a ruthless social system: as Henry James puts it in *Partial Portraits*, he is "a consummate picture of English brutality refined and distilled."[102] But he is, at the same time, a marginal case, an outlier and alien who frequently gets marked as such by the other characters. Duncan notes the "zoological language" that is often used by both narrator and character to describe Grandcourt: "developmentally arrested at the saurian stage, his coldblooded will to power unchecked by the gentle faculties of reason or sympathy."[103] These are not opposing positions; rather, taken together, they express the way the vicious spirit of the social and economic world (dehumanizing, extractive, imperial) is always simultaneously acknowledged and disowned by those benefiting from it. His arrival at Diplow provides a stimulant and center of gravity for the community's commercial forces, both in the usual sense of commodity exchange ("the corn-factors, the brewers, the horse-dealers, and saddlers, all held it a laudable thing, and one which was to be rejoiced in on abstract grounds") and through the spur it gives to the local marriage market.[104] In this "abstract" sense he functions as a typical gentleman, clad in all the obvious incentives for people to continue regarding him as such despite all of his cold-blooded peculiarities. Grandcourt's perfunctory, ironic observance of social niceties always threatens to give the game away: "What a washed-out piece of cambric Grandcourt is!" remarks Mr. Vandernoodt to Deronda. "But if he is a favourite of yours, I withdraw the remark."[105] As a result, he is accepted and rejected at once; the characters register a vague sense of discomfort with him as long as it can be done in a manner that comes at no social cost.

Mr. Gascoigne, in some ways the chief spokesperson and apologist for the prevailing order, illustrates the method by which Grandcourt's aberrant behavior keeps getting worked back into the fabric of the respectable social world:

> Whatever Grandcourt had done, he had not ruined himself; and it is well known that in gambling, for example, whether of the business or holiday sort, a man who has the strength of mind to leave off when he has only ruined others, is a reformed character. This is an illustration merely: Mr. Gascoigne had not heard that Grandcourt had been a gambler; and we can hardly pronounce him singular in feeling that a landed proprietor with a mixture of noble blood in his veins was not to be an object of suspicious inquiry like a reformed character who offers himself as your butler or footman. Reformation, where a man can afford to do without it, can hardly be other than genuine. Moreover, it was not certain on any other showing hitherto that Mr. Grandcourt had needed reformation more than other young men in the ripe youth of five-and-thirty; and, at any rate, the significance of what he had been must be determined by what he actually was.[106]

The pileup of excuses with their tendentious "moreovers" and "at any rates"; the permissive deferral of judgment; the strategic embrace of uncertainty ("whatever Grandcourt had done"); Gascoigne's wish, free indirectly rendered, is to know and not know about it. To be an operator in this world without ever having to look too closely at it. We can hear even in the repetition of the term "reformed character" the class-based inflections through which scrutiny is differentially applied and questions get begged: when "character" means "personal qualities," it is something that can be reformed; but when character means "role" or "part played," that reformation is something to be suspected as mere theater. Of course, and as usual, the blended quality of the discourse means that the narrator's archness is inseparable from Gascoigne's rationalizations, and we sense in places ("this is an illustration merely") that he is perhaps half-aware of the way his own exculpatory rhetoric keeps inadvertently inculpating Grandcourt. Later, we find him still busily talking himself into the match via free indirect discourse: "This was the view of practical wisdom; with reference to higher views, repentance had a supreme moral and religious value. There was every reason to believe that a woman of well-regulated mind would be happy with Grandcourt."[107] Gascoigne here is not simply excusing Grandcourt's past behavior; he is already scaffolding the victim-blaming narrative through which a future failed marriage can be understood. Any problems can be ascribed not to Grandcourt's cruelty or his lack of common human feeling but to Gwendolen's inability to "regulate" herself. But it is telling that Deronda will never need explained to him the reasons *why*, after just a few months of this marriage, Gwendolen has been rendered desolate and desperate and, finally, nearly homicidal. The reasons why Grandcourt would make someone feel those things have always been there to be known.

And indeed, one of the advantages of Gascoigne's half-knowing position, from his perspective, is that it allows him to claim the moral high ground no matter what actually happens. When the terms of Grandcourt's will are revealed after his death, with its generous provisions for his secret mistress and son, Gascoigne's response is a mix of outrage on Gwendolen's behalf and self-satisfaction that his suspicions have been confirmed:

> The Rector was deeply hurt, and remembered, more vividly than he had ever done before, how offensively proud and repelling the manners of the deceased had been towards him—remembered also that he himself, in that interesting period just before the arrival of the new occupant of Diplow, had received hints of former entangling dissipations, and an undue addiction to pleasure, though he had not foreseen that the pleasure which had probably, so to speak, been swept into private rubbish-heaps, would ever present itself as an array of live caterpillars, disastrous to the green meat of respectable people.[108]

No surprise, Gascoigne finds he has perfectly understood the situation the whole time and can thus deem his own behavior impeccable at every turn. It seems he had known just enough to be able to claim he had never been fooled by Grandcourt but never enough to recognize the threat he posed. Again, the question hinges on Grandcourt's typicality: Whereas a cultivated uncertainty about the man's past allowed for a "boys will be boys" kind of shrug ("it was not certain . . . that Mr. Grandcourt had needed reformation more than other young men in the ripe youth of five-and-thirty"), the will's stark acknowledgment of his "entanglements," its translation of his arrangement of personal loyalties into the language of money, combined, of course, with the very fact of his death, means he now can be posthumously evicted from the respectable world.[109]

Gascoigne's language here is still carefully euphemistic, expressing his "outrage" mostly at the violation of proprieties, suggesting he perhaps still fails to take Grandcourt's measure. All the more shocking, then, is the imagery of the final lines, like a sudden rip in the usual smooth rhetorical fabric that reveals an unaccommodated, unstable remainder of emotion, perception, imagination. The image of Lydia Glasher and her children as hungry insects hatching in the trash might be described as eco-gothic, merging, as it does, the familiar logic of that mode whereby buried sins inexorably return with a glimpse of the mindless processes of material nature. The classist disgust voiced here, with its hints of Malthusian and even eugenicist fears of overbreeding, suggests, I think, that we are meant to take this as Gascoigne's point of view. We might also hear in the hesitant "probably, so to speak" how this character must actively struggle to find appropriate language to manage this transgressive situation. The horror of the material inheres not simply in the idea of other people as blind consuming hordes but in the way their trespassing return triggers a more threatening kind of wholesale category breakdown. That is, "respectable people" are turned for a moment into "green meat"—vegetable matter made flesh; the ordinary plant-based diet of insects weirdly marbled with fat and thus larded with implications of cannibalism. The usual dehumanizing "othering" logic fails to work: there is an imaginative breach in the fictional *cordon sanitaire* ordinarily used to demarcate *them* from *us*. The Glashers could have been described as chewing their way through the social fabric or some other such metaphor; instead, we get a momentary glimpse of the entire human world as a maelstrom of indiscriminate feeding and breeding.

I know I'm belaboring this small moment, but the smallness is the point. Eliot's procedure here and elsewhere is to allow glimpses of something disquieting and even horrifying to show momentarily through the verbal surface that ordinarily keeps them out of sight. "Extermination," "to take all existence as fuel," "green meat"—such terms show alternative linguistic combinations, other strange realities that can flash into view the way the doors concealing the "death's head" sometimes unexpectedly pop open in the drawing room at Offendene. This, in part, is what Duncan refers to when he speaks of the "weird distortions of the novel's representational norms" in *Daniel Deronda*—the sense that some other, wider and more inhuman, reality continually threatens to erupt through the realistic surface.[110] He describes Gwendolen's unexpected "fits of spiritual dread"[111]—her deep-seated terror of isolated encounters with open natural spaces—in these terms: "It is as though the cosmic wilderness of organic life intuited by Dorothea in *Middlemarch* looms closer, larger, filling or rather emptying the sky."[112] These too are unexpected and momentary occurrences, ripples both within the diegetic world and across the narrative surface. But the unpredictability with which such realities can burst into awareness only attests to their lurking presence behind the forms of ordinary social life: "She was ashamed and frightened, as at what might happen again, in remembering her tremor on suddenly feeling herself alone, when, for example, she was walking without companionship and there came some rapid change in the light. Solitude in any wide scene impressed her with an undefined feeling of immeasurable existence aloof from her, in the midst of which she was helplessly incapable of asserting herself."[113]

The ubiquity of this empty reality seems to call forth in Gwendolen's mind a kind of resistant counterpressure. Both her mental habits and her story alike are characterized by denial. But Eliot is careful to show us that this denialism is not merely a quirk of this one individual temperament facing down the universe but the product of an ongoing dynamic between her mind and the social world that has set the terms through which she perceives herself. Her sense of identity, Eliot shows, is made up of a thicket of contradictions by which certain socially agreed-upon constructions of womanhood keep her actual situation concealed in superficial and temporary ways. When the novel begins, it is clear that whatever "power" she has inheres only in her ability to select the man to whom she will surrender all of her power, and thus her much-discussed individual "superiority" makes her actual position of structural inferiority easier, for a time, to forget.[114] But such a habit of willful forgetting also perilously widens the gap between the make-believe and the real, rendering the latter increasingly unfamiliar and harder to countenance, which thus incentivizes a deeper attachment to the former. Her spells of agoraphobia make the stress and extremity of her denialism plain to see—the silence of isolation reveals the impossible load-bearing function that the airy fabrications of social rhetoric have taken on. But such moments don't simply expose Gwendolen's idea of self-sovereignty as a known fiction; they also suggest that the very processes that make that fiction possible—the mechanisms of denial itself—reveal a deeper self-alienation, a lack of control over herself via the very means that allow her to believe in it. The narrator remarks of her: "There is a great deal of unmapped country within us which would have to be taken into account in an explanation of our gusts and storms."[115]

Such a moment could be (and has been) productively read as Eliot's account of Gwendolen's unconscious mind: Louise Penner, for example, sees this moment as an "invitation to interpret the cause of Gwendolen's symptoms of psychic trauma," which she reads as possibly the "repressed memory of childhood incest."[116] And Carole Stone, in a similar vein, proposes that Gwendolen's psyche is troubled, even defined, by repressed memories of her father.[117] Intriguing as these symptomatic readings are, I'm more interested in the way in which the character seems, on some level, not only aware of upsetting or aversive realities but also aware of how she is continually working to push them out of mind. When she speaks one-on-one with her uncle, she seems to recognize that his conversation makes certain pleasing fictions seem more real: "This evening she was willing, if it were possible, to be a little fortified against her troublesome self. The Rector's mode of speech always conveyed a thrill of authority";[118] "she felt as if she were reinforcing herself by speaking with this decisiveness to her uncle."[119] Later, she explicitly acknowledges her tendency to reality management: "I wish I had never known it!" she says, referring to the information about Grandcourt's secret family.[120] Indeed, it is the Glasher subplot that provides the clearest example of Gwendolen's denialism, the reality that she always knows and does not know.

Thus, Gwendolen spends the early part of the novel in such a state of precarious self-consciousness in which she desperately wants to believe in the polite fictions constructed around her, while she also remains intelligent enough—and is enough of an outsider to this social world—to recognize the game for what it is. "How you treat us poor devils of men!," Grandcourt tells her. "'We are always getting the worst of it.' '*Are* you?' said Gwendolen in a tone of inquiry, looking at him more naïvely than usual. She longed to believe this commonplace *badinage* as the serious truth about her lover."[121] Longs to, but can't; and, on some level, knows it. We see—and she does too—the poison lacing the proffered apple. If only she could picture Grandcourt as a victim, it would let her continue indulging in the fiction of her power over men, and it would also allow her to dismiss Lydia Glasher as a seductress who lacks any serious moral claims upon her. This is what Cohen would call "interpretive denial," and the temptations of such a configuration are obvious; but "longing" to believe something is clearly not the same as actually believing it.[122] Instead, it suggests a meta-awareness of the motivated process attending belief itself, her understanding of the larger culture of denial in which she is being asked to participate. Gwendolen cannot make herself believe in Grandcourt's victimhood, not only because it's conspicuously untrue but also because she recognizes how much depends on the mystifying power of social rhetoric. It is not just that such language is unmistakably insincere and flimsy, but language *itself* is untrustworthy—too subject to reversal and revision. It continually offers itself as the ready-to-hand building materials through which one can construct a pleasing reality to house a desired sense of self, but it therefore always makes that reality and sense of self vulnerable to the next thing out of someone's mouth: "When you talk in that way my life shrivels up before me," she snaps at her mother, after the latter makes a stray remark about aging.[123] As one of the Meyrick children says later in a different context, "We must have the courage to hear things, else there is hardly anything we can talk about."[124] As the chief instrument of denial, language is also always the means of its undoing.

The ambiguity attending Gwendolen's "movement of mind" and the traffic between various states of self-awareness is consistently rendered via free indirect discourse.[125] At times, such voicing ironically exposes the defensive properties of language; at other times, it allows for more disturbing possibilities to enter the ambit of awareness. After first imagining how suitably emotionless Grandcourt seems (and thus, she imagines, how pliable a partner he'll make), Gwendolen's mind free indirectly drifts toward more unsettling implications:

> How was it that he caused her unusual constraint now?—that she was less daring and playful in her talk with him than with any other admirer she had known? That absence of demonstrativeness which she was glad of, acted as a charm in more senses than one, and was slightly benumbing. Grandcourt after all was formidable—a handsome lizard of a hitherto unknown species, not of the lively, darting kind. But Gwendolen knew hardly anything about lizards, and ignorance gives one a large range of probabilities.[126]

Those first two questions are clearly Gwendolen's, and they convey both her sense of Grandcourt's distinction and her misgivings about the match. But how to categorize what follows? Is the word "benumbing" the narrator's or hers? How aware is she of the hypnotic effect that his remote manner has upon her? Does she actually think of him as a "handsome lizard," or is this merely the narrator's vivid way of rendering her vague sense of him as alien, unfeeling, cold-blooded? Does that last sentence show her deliberately walling off further inquiry? *Daniel Deronda* constantly shifts into and out of free indirect discourse in this way, and because the transitions are never clearly marked, such unresolvable meta-questions keep arising—are, in fact, constitutive of the representation of her consciousness. If we ask, Does Gwendolen on some level *know* that Grandcourt is threateningly reptilian?, the answer seems to be that she both does and does not. As the last sentence of the passage suggests, her ignorance has something strategic about it, for it allows her to continue to feel as if she takes the man's measure without having to foreclose the possibility of a socially advantageous match. The mix of clear-sighted understanding and wishful thinking, of simultaneous noticing and ignoring, allows her to maintain, for the moment, a fiction of comprehensive superiority.

This nebulous in-between state of knowing and not knowing expresses itself in the indecisiveness and passivity that characterize the important decisions she makes in the novel. First, she both does and does not consent to Grandcourt's marriage proposal—as F. R. Leavis says, "No acquiescence could look less like an expression of free choice."[127] She shies away from making up her mind, offering a plain verbal affirmation, until the circumstances engineered by Grandcourt bring things to a crisis:

> "Do you command me to go?" No familiar spirit could have suggested to him more effective words.
>
> "No," said Gwendolen. She could not let him go: that negative was a clutch. . . .
>
> "You accept my devotion?" said Grandcourt, holding his hat by his side and looking straight into her eyes, without other movement. Their eyes meeting in that way seemed to allow any length of pause, but wait as long as she would, how could she contradict herself? What had she detained him for? He had shut out any explanation.

"Yes," came as gravely from Gwendolen's lips as if she had been answering to her name in a court of justice.[128]

Realistically depicted as he is, Grandcourt functions in the allegorical scheme of the novel as Gwendolen's nemesis figure. "His long narrow grey eyes expressed nothing but indifference": it is as if the indifference of the world that has long terrified her has assumed human form and come to reside with her in horrifying intimacy.[129] The very social rhetoric that Gwendolen has previously used to fortify herself has, in this scene, become a set of walls put up one by one, in front of our eyes as it were, until she is chained within the very forms in which she once took refuge. Grandcourt is both plausibly expert at this kind of verbal gamesmanship and almost demonically gifted at sounding out Gwendolen's weaknesses. An utterance now actually does cause a reality to come into being, insofar as the simple speech act of voicing a "yes" to his question—irrespective of the clarity of intention behind the utterance—will literally create the conditions in which the rest of her life will unfold.

Near the end of the novel, in a marvelous bit of symmetry, she escapes her marriage in a similarly ambiguous and uncommitted manner, insofar as she both does and does not kill her husband. When she hesitates to throw Grandcourt the rope to save him from drowning, it is as if she is allowing her murderous desires to find shape and expression in the world. She tells Deronda, "I held my hand, and my heart said, 'Die!'—and he sank; and I felt 'It is done—I am wicked, I am lost!' "[130] The uncanny connection Gwendolen has always felt between vocalizing something and making that thing real in the world here finds grotesque expression; in this case, the imperative "die" is spoken twice, once in a silent inner wish and once fully aloud in the retelling. The repetition is crucial, because it suggests she is misaligned now in the other direction: that is, instead of being afraid to vocalize certain unpalatable realities, Gwendolen is (to use a gambling metaphor befitting this novel) doubling down on them. She now fully affirms the aversive vision of herself as adrift in a void, cut off from others by her own hidden moral deficiencies. Of course, this is too extreme as well, which Deronda plainly recognizes. The idea that one can speak things into reality continues to imply an unwarrantably grand sense of sovereignty that, although putatively put in the service of atonement here, commits her to a similar moral error. But Gwendolen is not

Bulstrode, and she will not be left stranded in this state of denial. Although, like him, she has encountered her nemesis and wishfully, half-willingly participated in his death, and although, also like Bulstrode, her own pressurized moral isolation warps her view of reality into a psycho-drama with herself at center stage (she imagines her courtship and marriage to Grandcourt as "a long Satanic masquerade, which she had entered on with an intoxicated belief in its disguises, and had seen the end of in shrieking fear"), there is a path out of illusion for her.[131] That path is lighted by Deronda, of course, both in his role as confessor and instructor and through the profound shock his own narrative delivers to her self-centered organization of reality. When he informs Gwendolen of his plans to marry Mirah and embark upon his Zionist mission, she sees him at last not as a player in her private drama of sin and redemption but as the protagonist of his own: "That was the sort of crisis which was at this moment beginning in Gwendolen's small life: she was for the first time feeling the pressure of a vast mysterious movement, for the first time being dislodged from her supremacy in her own world, and getting a sense that her horizon was but a dipping onward of an existence with which her own was revolving. All the troubles of her wifehood and widowhood had still left her with the implicit impression which had accompanied her from childhood, that whatever surrounded her was specially for her."[132] In this way, Deronda ironically completes the humbling Gwendolen experienced at the hands of Grandcourt. It's doubly ironic, in fact, because it both suggests a kind of functional continuity between the two men where she had previously only registered difference and because that continuity emerges only from her understanding that his life cannot be reduced to its function in hers.

Deronda's plot is also characterized by denial, but of a slightly different variety. Not subject to fits of existential dread or fears of his own insignificance as Gwendolen is, Deronda avoids what is potentially unsettling knowledge about his racial identity. But though, like Gwendolen, he lingers through the early parts of the novel in a nebulous zone of inaction and indecision, knowing and not knowing, he takes refuge not so much in self-protective social fictions but in a strategic blurriness of motivation. His search for Mirah's family proceeds in a strangely but expressively ambiguous manner in which he seems to be both looking and not looking for them. The reasons for his half-heartedness are various: it seems partly because he is a snob and maybe even a bit of an anti-Semite, filled, as he admits, with a "sense of repulsion at the commonness of these people."[133] If Mirah is related to people as stereotypically vulgar (in his mind) as the Cohens, he does not want to know it. But it's also because he is attracted to her, which he does not want to admit, but which might become harder to avoid once he restores her to her biological family and surrenders his role of disinterested guardian. It seems possible too, that, like the Meyricks, he secretly hopes that if she does not reconnect with her family, she will eventually assimilate or even convert and (perhaps) thus become an eligible marriage partner for a putatively Christian gentleman. On yet another level, he seems to see in her story a version of his own, and once she reconnects with her own heritage, the way would suddenly be clear for him to do the same: "His conscience was not quite easy in this desire for delay, any more than it was quite easy in his not attempting to learn the truth about his own mother: in both cases he felt that there might be an unfulfilled duty to a parent, but in both cases there was an overpowering repugnance to the possible truth."[134] All of these motivations (and others) are present in varying shades of dimness and clarity in his mind and of course exist in relationships of tension and even contradiction with each other. He does and does not want to view Mirah as a lover; he does and does not want her to be Jewish; he does and does not want to find he is Jewish himself. These sets of unresolved binaries are interarticulated with each other insofar as different combinations make different future realities possible and impossible.

But not one of these motivations is ever fully owned, and that lack of ownership is again conveyed via Eliot's distinctive use of free indirect discourse in which the distinction between narratorial probing and a character's own reflective self-scrutiny is left strategically vague. Deronda's confrontation with his own motivations—his desires, his aversions, his unexamined prejudices—can be deferred through the mechanism of a perpetually unsolved quest, and on some level, he realizes this is the case:

> Why did he not address himself to an influential Rabbi or other member of a Jewish community, to consult on the chances of finding a mother named Cohen, with a son named Ezra, and a lost daughter named Mirah? He thought of doing so—after Christmas. The fact was, notwithstanding all his sense of poetry in common things, Deronda, where a keen personal interest

was aroused, could not, more than the rest of us, continuously escape suffering from the pressure of that hard unaccommodating Actual, which has never consulted our taste and is entirely unselect.[135]

We hear Deronda's voice in that opening question and perhaps also in the interruptive "—after Christmas," which perhaps signals how his self-interrogation gets habitually and formulaically postponed. The last sentence has more of the ring of narratorial distance, but it also poses a telling dilemma: Does the "hard unaccommodating Actual" mean the life of the "common Jews" whom Deronda observes in his ramblings in St. Mary Axe and Whitechapel?[136] Or does it mean the idea of Mirah's connection to them? Is his inability to "continuously escape suffering" the recognition that he cannot put off the discovery forever, or is it his present experience of imagining Mirah to be connected to "the dingy shops and unbeautiful faces" he passes?[137] It could be either. But the difference is crucial, because it signals not simply a personal crossroads for Deronda but also a narrative crossroads for Eliot. For Mirah to turn out to be a member of the "common" shopkeeping lot is the path of realism—not because of her class status as such but because of the way her "commonness" would disrupt Deronda's desire for the world to match his ideal imaginings. This is Eliot's familiar method of establishing a sense of realism: a character must come to terms with the gap between his or her own personal construction of reality and the hard fact that reality is an impersonal and unyielding thing. But such a plot of disillusionment, adumbrated as it is in this scene, is not what happens—Mirah, we learn some hundred sixty pages later, is not part of the common crowd and Deronda's suffering is thus located in his *deferral* of the discovery, not in the discovery itself. The Actual turns out to be entirely accommodating and suited to his "taste." Since we don't realize that until much later, however, the ambiguity in the description of Deronda's "suffering" here is a bit of misdirection, where Eliot appears to be subjecting his story to the same disenchanting rigors as Gwendolen's but is in fact, at the same moment, preparing us for something like the opposite.

As we have seen so often, Eliot is concerned with the way language works to conceal and reveal simultaneously, showing how the rationalizing defenses of her characters often wind up inadvertently expressing something of the problematic reality they seek to manage. Something similar is happening here, only with the polarity reversed: the narrator appears to be working on the side of disenchantment and disillusionment but is actually, *and by way of the same linguistic construction*, covertly setting the stage for a re-enchantment. The language mimics the process described at the end of the paragraph, where signs become legible either prophetically or retrospectively through the contextualizing frame of historical perspective:

> Here undoubtedly lies the chief poetic energy:—in the force of imagination that pierces or exalts the solid fact, instead of floating among cloud-pictures. To glory in a prophetic vision of knowledge covering the earth, is an easier exercise of believing imagination than to see its beginning in newspaper placards, staring at you from the bridge beyond the corn-fields; and it might well happen to most of us dainty people that we were in the thick of the battle of Armageddon without being aware of anything more than the annoyance of a little explosive smoke and struggling on the ground immediately about us.[138]

As in *Middlemarch*, the movement of history is imagined as the organizing principle that will rescue ordinary life from the chaos of mere materiality As in *Middlemarch*, energy is invoked as the mediator between the lived texture of the everyday and the unfolding of some grander story that marshals and conserves what might seem in the moment, or from a limited vantage point, to be diffusive, wasteful, and pointless (here, struggle and smoke). The difference is that in *Daniel Deronda* she leans much further into a vision of history as somehow (the ambiguity is key) spiritually and racially organized rather than just vaguely progressive. The idea of a "prophetic vision" unfolding in time is offered provisionally, as a thought experiment rather than a positive attestation, and thus it comes (here as elsewhere) with a measure of plausible deniability. We needn't entirely agree with Leavis's wholesale dismissal of the novel's "Jewish plot" to recognize that it offers a resolution that Eliot herself does not entirely own: "There is no equivalent of Zionism for Gwendolen, and even if there were—: the religion of heredity or race is not, as a generalizable solution of the problem, one that George Eliot herself, directly challenged, could have stood by. . . . All in the book that issues from this inspiration is unreal and impotently wordy."[139]

The significance of ethnic nationalism in *Daniel Deronda* has been much debated in scholarship on the novel, with critics like Amanda Anderson,

Kwame Anthony Appiah, and Thomas Albrecht finding in Daniel's Zionism something much more flexible and compatible with the universalist ethic of liberal cosmopolitanism than perhaps at first appears. Anderson writes: "Through the character of Deronda, Eliot advocates a form of cultural self-understanding that might be best called reflective dialogism: her model for one's relation to history, culture, and nationality becomes passionate argumentation, not simple embrace. To achieve reflective distance, one must be capable of disengagement from cultural norms and givens. But such achieved distance should in turn promote not a sustained or absolute disengagement—for Eliot a destructive delusion—but rather a cultivated partiality, a reflective return to the cultural origins that one can no longer inhabit in any unthinking manner."[140] Anderson is surely right that Deronda's vision of ethnic nationalism is meant to be seen as significantly more complex and sophisticated than Mordecai's troublingly Romantic, and un-self-critical vision of an organically unfolding racial destiny. And yet it's hard not to also feel that this very complexity and sophistication, preferable as they may be to what Mordecai offers, represent their own kind of magical thinking, this time of the hyper-literate and nuanced variety. The idea that there's some way to synthesize the good parts of liberal cosmopolitanism and racial ethno-nationalism if only one is able to reflect on it critically and "dialogically" enough remains frustratingly, almost definitionally, vague. Although Anderson is careful at the end of her chapter to qualify this valorization of critical self-reflection—"even the most reflective dialogical models of cultivated detachment can harbor violent blindnesses and exclusions"—one can't help feeling that self-consciousness is nevertheless being fetishized, especially its ability to qualify and reflect critically upon itself.[141]

Aleksandar Stević argues that readings such as Anderson's and Appiah's work by redefining "cosmopolitanism" so capaciously as to threaten to empty it of meaning.[142] So what the synthesis between these paradigms would actually look like in practice and how exactly the logic of racial exclusivity is to be wedded to a liberal ethic of inclusivity are matters always to be deferred. As Anderson puts it, "The precise form of his cultural identification will grow out of reflective judgment, selective appropriation, and a process of dialogue between himself and others."[143] In this sense, the disavowal that characterized Deronda's initial reluctance to discover his ethnic identity is not overcome at all but is simply incorporated into his relation to that fact, turned into a positive virtue. We are back to the dynamics of the veil and the way reestablishing a mysterious just-out-of-reach resolution substitutes form for content. Open-endedness becomes its own end: the solution offered in *Daniel Deronda* is the belief that there will be a solution.

As Stević points out, in the one moment when there is a choice to be made between these competing paradigms, Deronda decides to part ways with Gwendolen: "She (and the reader, for that matter) is expected to accept that blood is destiny, that ethnic origin is the primary and decisive marker of identity which easily overrides other commitments, and, finally, that the man in whom she confided has become a stranger simply by virtue of discovering that he is an ethnic Jew. She is not asked to embrace an ethics of alterity but to come to terms with what the novel posits as the natural priority of ethnic bonds. The imperative of separateness in *Daniel Deronda* is inseparable from the logic of nationalism."[144] The separation effected along these lines is also, at heart, a separation between the "indifferent if not meaningless existence" that characterizes much of the Gwendolen "half" of the novel and the organized, purpose-filled life that Deronda achieves with the discovery of his Jewish identity. Stević argues that Eliot connects this threat of meaninglessness with a cosmopolitanism that is unrooted in the land and unconnected to national tradition, vividly illustrated by the vacant crowd of casino-goers at Leubronn, among whom Deronda first spots Gwendolen.[145] The casino is, I would add, conspicuously defined by the blind operations of chance, which suggests a concomitant moral chaos from which Deronda's decidedly teleological plot (the historical one he imagines, as well as the personal one he inhabits) also seems to cordon itself off.

What we find in *Daniel Deronda*, then, is a complex and multilayered version of the kind of disavowal that so often characterized the response to Darwin's undirected, chance-driven picture of the natural world. What appears to rescue the human from this chaos, from the frightening vision of the "amalgamation with matter," is a racialized understanding of history. In this case, Eliot's decision to focus on Jewish national identity allows her to avoid—and, indeed, even ironically upend—the pernicious white supremacism of British imperial discourse and its attempts to reestablish a hierarchy of humanness on racial grounds. And yet, Eliot's solution still partakes of the same toxic logic, including the same fetishization of self-consciousness as both illustration of, and means of realizing, some kind of "higher" spiritual, maybe even non-animal, existence. How toxic it is can be quite clearly seen in the manifold horrors committed in the name of Zionism in the twentieth and twenty-first centuries. As Duncan notes, Gwendolen and Grandcourt are consistently given "reptilian traits" and described in "zoological language, with its Darwinian cladding"; Deronda, meanwhile, "in contrast to these primitives, embodies a future human type around whom the decadent formations of national life may be reintegrated."[146] Now, is Eliot actually saying that there is some qualitative, biological difference between Daniel and Grandcourt? Of course not. All the same, she's not *not* saying it either. Similarly, the fact that Mordecai's prophetic claims are all proven correct lends them a sheen of validity, even as we know we are not *really* meant to believe in his mysticism. In this way, the disavowal of Darwin occurs by means of the very conventions of the novel, in the figurative language and plot devices that are taken as real even as they are not taken seriously.

FIVE

VIRGINIA WOOLF AND THE ENDS OF DENIAL

IN HER 1926 ESSAY *On Being Ill*, Virginia Woolf discusses the experience of illness, writing not so much about the suffering of the sickroom but about the forms of knowledge the sick person gains—or regains—access to:

> It is only the recumbent who know what, after all, nature is at no pains to conceal—that she in the end will conquer; the heat will leave the world; stiff with frost we shall cease to drag our feet about the fields; ice will lie thick upon factory and engine; the sun will go out.[1]

The idea of irreversible global energy loss and the eventual so-called heat death of the sun was not only a much-debated scientific question in the second half of the nineteenth century (courtesy of the nascent field of thermodynamics); it was a spur to all manner of end-of-days anxieties in novels and poems by Charles Dickens (*Bleak House*), H. G. Wells (*The Time Machine*), Camille Flammarion (*Omega: The Last Days of the World*), James Thomson (*The City of Dreadful Night*), and many others.[2] But Woolf here minimizes both the science and the apocalyptic narratives emerging from them by suggesting that such knowledge is available to anyone with a cold. Just as Darwin argues that the ruthless violence of the natural world is something we all already know but tend to "forget," Woolf posits that the entropic direction of everything is first and foremost a bodily experience, one that routinely gets ignored in the course of our "ordinary," ambulatory lives.

Or better yet: *somnambulatory*, since here, as throughout the essay, it is the ill who appear most fully "awake" to the truth of human experience, whereas those busy pursuing their everyday routines do so in massed, unthinking formations: "The army of the upright marches to battle. Mrs. Jones catches her train. Mr. Smith mends his motor. The cows are driven home to be milked. Men thatch the roof. The dogs bark. The rooks, rising in a net, fall in a net upon the elm trees. The wave of life flings itself out indefatigably."[3] The active voice yields to the passive as various "drivers"—of cars, of cattle—are themselves, we see, being driven by railway schedules and the demands of their machines, as the motions of the human world flow into the non-human animal and then, further, into the inorganic, the elemental, as life becomes a wave. Carried across the passage too, quietly, though unmistakably, is the force of Woolf's social critique. It is, she suggests, by submitting to the habitual routines of "civilized" life that a person becomes, perversely enough, most like an animal: unreflecting, pattern-driven, swarming in self-organizing nets. Conversely, and counterintuitively, it is the sick person, confined to bed and made undeniably aware of being "fastened to a dying animal" as Yeats put it around this time, who realizes what is most "human" about herself.

This is a familiar enough kind of modernist inversion, tinged with an also-familiar disdain for mass society and the goings-on of "ordinary" life. What I want to stress here is the way that Woolf positions herself with respect to an earlier (Victorian) generation of scientific naturalists. The signal discoveries of nineteenth-century science—the inevitable entropic direction of all things; the total continuity between human and non-human realms; the inevitability of personal and species-level extinction; the human-eclipsing scale of the natural world—are, Woolf suggests, totally obvious. Such things are simply there to be seen and experienced, because nature has taken "no pains to conceal" them. And thus, the only reason such basic facts can even seem like "discoveries" at all is that they've been concealed by other means and for other ends: by, for example, the mystifications of Christianity or—more to the point here—by the various, affiliated secular ideologies of progress and capitalist rationalization, especially those providing moral cover for

Great Britain's imperial designs. From start to finish, *On Being Ill* describes and at times, enacts, the ways in which these ideologies downplay or dismiss the basic reality of the animal body; they do so, it is suggested, because that body stubbornly belies the various fictions of transcendence and mastery entailed in the idea of "civilization" and because the body's various demands and failures are just so many inefficiencies within an inhumanly productivist economic regime. Woolf archly describes the sickroom as a place not only of bodily wasting but also of wasteful activity on the part of the nurses and caregivers who devote themselves to comforting their charges: "Sympathy nowadays is dispensed chiefly by the laggards and failures, women for the most part (in whom the obsolete exists so strangely side by side with anarchy and newness), who, having dropped out of the race, have time to spend upon fantastic and unprofitable excursions."[4] These "laggards" include "A. R., the rash, the magnanimous, who if you fancied a giant tortoise to solace you, or a theorbo to cheer you would ransack the markets of London and procure them somehow" and "the frivolous K. T., dressed in silks and feathers, painted and powdered (which takes time too) as if for a banquet of kings and queens, who spends her whole brightness in the gloom of the sick room, and makes the medicine bottles ring and the flames shoot up with her gossip and her mimicry."[5] Here and elsewhere, the voice is riven with tensions and internal conflicts: the descriptions of A. R. and K. T. are at once vivid and vague, descriptively filigreed and flatly anonymous. The mock-heroic rhetoric describing quixotic grail-knights and pretend royalty does its intended work, but it also strangely folds back on itself to suggest a picture of actual heroism, of gestures of sympathy admirable precisely *for* their extravagant ineffectuality.

On Being Ill, then, is an essay about denial—the everyday denial of the animal body and thus the denial of feeling, of compassion, of knowledge, of imagination. Of life itself, in a way. But it's also an essay that vocalizes denial through the tonal and generic instabilities of a voice at war with itself. Thus, different attitudes, positions, and moods emerge, clash, change, disappear, and reshuffle themselves, sometimes over the course of a few sentences:

> But in health the genial pretence must be kept up and the effort renewed—to communicate, to civilise, to share, to cultivate the desert, educate the native, to work by day together and by night to sport. In illness this make-believe

ceases. Directly the bed is called for, or, sunk deep among pillows in one chair, we raise our feet even an inch above the ground on another, we cease to be soldiers in the army of the upright; we become deserters. They march to battle. We float with the sticks on the stream; helter skelter with the dead leaves on the lawn, irresponsible and disinterested and able, perhaps for the first time for years, to look round, to look up—to look, for example, at the sky.[6]

The language of social doxa, including the imperatives of empire, is at first preserved, even as the whole project is acknowledged to be all pretense and "make-believe"; the morally loaded categories (upright vs. deserters) also remain in place, even as the voice puts them into new configurations: "disinterested," that quintessentially Victorian aspiration now, perversely enough, available only to the "irresponsible." By the end, the categories have been sufficiently remixed so that it is the dead, the inanimate, the chaotic that seem most alive, awake, and free; and the outcome is something like simple awareness, mere being. The key point is that it is by way of this mingled and shifting polyvocality, the sense that the resistant spirit of this essay is also tangled up in or in some ways constituted by, the discursive formations of a reductive social order, that *On Being Ill* imagines denial not as an individual failing but as a structure of feeling endemic to the culture at large. Critique and complicity tumble together in a single stream, and they'll keep tumbling, as we'll see, through the pages of her fiction.

And Darwin is at the center of this. In what follows, I read Woolf as the great diagnostician of denial, someone who observed critically, but not unsympathetically, her culture's inability or unwillingness to fully countenance the implications of its own scientific discoveries. Much of Woolf's work, but especially *To the Lighthouse*, is concerned with the abiding legacy of the nineteenth century and its culture of denialism for those living in its wake. As we have seen, the Victorians were aware of their own habits of willful forgetting and evasion, but such habits were often imagined as necessary, even psychologically healthy, defense mechanisms. What, after all, is the point of dwelling on natural inevitabilities—the eventual extinction of the human species or the heat death of the sun—or what is to be gained from imagining the human animal as simply one animal species among thousands of others? If human meaning and value were, on some level, acknowledged fictions, wasn't subscribing to such fictions preferable to simply surrendering those meanings and values to a chaotic, violent, and utterly pointless natural order? We have seen Eliot, Huxley, Wells, and even Darwin himself feeling the emotional, intellectual, and moral pull of denialism. And we have seen Conrad put the matter a good bit more cynically, imagining the denial of imperial realities as a widespread moral problem, a "nightmare" even, and yet also noticing that no apparent consequences trouble those in the metropole who have the luxury of encasing themselves in these fictions: "The heavens do not fall for such a trifle," Marlow remarks after lying to the Intended at the end of *Heart of Darkness*.[7] As Darwin's work made clear, there is no moral structure to things, no superintending providence, or economy of cosmic justice. Therefore, it is simply not the case that those who decide to turn a blind eye to their own participation in a regime of horror will ever receive their comeuppance or even realize what it is they are actually a part of.

Woolf dissents in an intellectually daring and formally innovative way from both of these positions. For one, she isn't so sure that denialism is psychologically necessary or healthy, though she does recognize the intellectual, affective, and psychological difficulties involved in figuring out what and how much of reality one can reasonably absorb. What must be kept always in mind; where the cut gets made between the included and the excluded; where the foreground can wash away into the background; what and how much should be remembered of what seems to want to be forgotten—these moral and epistemological questions, which many of her characters (Lily Briscoe, Rachel Vinrace, Clarissa Dalloway) must work out for themselves, are also rigorously pursued on the level of narrative form. The problem of denialism—especially denial of the non-human realm—is a consistent topic of interest in her fiction and essays that becomes a key aspect of her reimagining of literary realism and thus part of the experience of reading her books. Woolf's interest in the unceasing, moment-by-moment negotiation between the perceiving, remembering, embodied mind and its surroundings—with what she calls the "myriad impressions—trivial, fantastic, evanescent, or engraved with the sharpness of steel" that come at the perceiver "from all sides . . . an incessant shower of innumerable atoms"—means that denialism is imagined less as a fixed condition or character flaw, more as an ongoing process by which non-human realities are known and not known, attended to and ignored, confronted and sidestepped, remembered and forgotten, all in ceaseless, shifting, kaleidoscopic patterns of thought and feeling, informed by past history and present circumstance, generated by what is felt by the body, understood by the mind, changed and distorted through language, and falsified by social conventions and forms.[8] Characters are not so much "in denial" but in and out of it, constantly. Or perhaps better (if more awkward) to say they are "in disavowal," insofar as that term, as we have seen, describes a continuous, unsettled push and pull taking place in time. Her novels throw upon the reader the question of how, or if, a coherent picture can be made out of these shifting patterns, thereby forcing an awareness of what inevitably gets sacrificed in the name of coherence. In this way, denial, or disavowal, is encoded into narrative form, as acts of willful exclusion become problems for the reader to manage.

Such issues can be seen in all of her fiction, but denialism is most clearly centered as topic, structure of feeling, and narrative dynamic in *To the Lighthouse*, a novel that (not coincidentally) involves her most thorough critique of nineteenth-century culture and its aftermath. What that novel shows is that a culture of denialism, far from making possible necessary psychological defenses, actually exacts profound, hidden costs, especially from the women who make certain fabrications and forms of forgetting possible. Further, such a culture produces wider global and we might even (advisedly) say *ecological* consequences that leave no one unscathed. The heavens actually *do* fall in *To the Lighthouse*, in a sense, in the form of the Great War, a catastrophe that bursts into view in the middle section of the novel and seems connected in some undefined way to the kinds of evasions we see organizing the Ramsay household in the first section. The *fact* of the connection is there to be inferred, even if the actual chain of cause and effect is left suggestive, implicit, indeterminate. In this way, I argue, Woolf suggests something of the uncertain scalar logic of the climate crisis, in which tiny individual behaviors, informed by more widespread cultural fictions, are connected in obscure but decidedly material ways to larger global phenomena. The war emerges not out of some providential principle of moral scale-balancing but from the inevitable breakdown of the unsustainable and obfuscatory myths of empire and capital. In this way, the Great War is a key juncture in environmental history not simply because of its devastating ecological consequences, or its origins in extractivist logic and the scramble for global fossil-fuel or mineral reserves, but because, as Bruno Latour argues, it shows the entire world "sleepwalking" into a catastrophe that was entirely predictable and avoidable, one that featured innumerable obvious warning signs and exit ramps that were both noticed and disregarded.[9] What *To the Lighthouse* does so brilliantly, I argue, is show how this catastrophe was always hiding in plain sight so that when it finally bursts into the diegetic world of the narrative, it comes with a strangely mingled sense of shock and inevitability. The reader is surprised by things she already knows, making clear, retrospectively, the work of a denialism that seemed to have been safely satirized, identified, and overcome. The novel thus dramatizes the transition beyond the Victorian-era denial or disavowal of non-human realities, to what Timothy Morton describes as facing the profound "difficulty" of ecological thought: "becoming open, radically open—open forever, without the possibility of closing again."[10]

WOOLF AND DARWIN

Although *On Being Ill* might cheekily downplay the scientific discoveries of the previous century, Woolf's quarrel is not so much with the content of that work as with the way it had been mythologized, distorted, selectively invoked, and pressed into service for various ideological ends. As Gillian Beer has shown, Victorian scientific naturalism was foundational for Woolf's development as a thinker and artist, especially the work of that "major figure in her own upbringing, her own 'development,' Charles Darwin."[11] Her first novel, *The Voyage Out* (1915), bears a striking and obvious debt to Darwin's *Journal of Researches* (aka *The Voyage of the Beagle*), which provided her with detailed descriptions of the tropical forests of South America and made it possible for her to set a novel in a place she had never actually been. But Darwin's influence on Woolf goes much further and deeper than that. As Beer has persuasively argued, his work informed her understanding of everything from heredity to history to free will to temporality to the definition of the human. Indeed, her final novel, *Between the Acts* (1941), with its many extended discussions of prehistory, Deep Time, and human evolution, finds Woolf still actively working through Darwinian ideas near the end of her life. One can clearly see in that text how those ideas retained their strange power to shock and unsettle, when, in the opening, a character called Mrs. Swithin is imaginatively upended by a book on natural history:

> [She] had spent the hours between three and five thinking of rhododendron forests in Piccadilly; when the entire continent, not then, she understood, divided by a channel, was all one; populated, she understood, by elephant-bodied, seal-necked, heaving, surging, slowly writhing, and, she supposed, barking monsters; the iguanodon, the mammoth, and the mastodon; from whom presumably, she thought, jerking the window open, we descend.[12]

The point is not that such ideas are new or groundbreaking; it's that they aren't. The natural history underpinning Darwinian evolutionary biology is familiar and ready to hand (we're told this is her "favourite reading") and yet also somehow still disorienting. As always with Woolf, the real work happens through the form, in the way the sentence returns obsessively to the notation of Mrs. Swithin's mental acts—"she understood . . . she understood . . . she supposed . . . she thought"—marking and marking again the gap between knowledge and knower and growing more hesitant as she approaches the question of human evolution. As Mrs. Swithin's mind halts and pauses over imagined vistas, her actual body performs in a smoothly habitual manner the usual motions of rising from bed, moving to the window, opening it. Opposing scales and dimensions—the immediate and the abstract, the embodied and the imagined, the time of morning ritual and the time of geological unfolding—may be housed uneasily together in one person's mind, but they remain conspicuously unintegrated, punctuated at last by the contrasting movement of a window rising while the human race "descends." So confusing is all this that it takes Mrs. Swithin a moment to realize that when the door opens and someone enters her room, it is not a "beast in a swamp" but her maid Grace bringing breakfast.[13]

In *Mrs. Dalloway* (1925), we learn, via Peter Walsh, that Clarissa Dalloway spent her childhood reading the works of Victorian scientific naturalists and that this experience has left her dwelling in a set of paradoxes:

> As we are a doomed race, chained to a sinking ship (her favourite reading as a girl was Huxley and Tyndall, and they were fond of these nautical met-

aphors), as the whole thing is a bad joke, let us, at any rate, do our part; mitigate the sufferings of our fellow prisoners (Huxley again); decorate the dungeon with flowers and air cushions; be as decent as we possibly can. Those ruffians, the Gods, shan't have it all their own way,—her notion being that the Gods, who never lost a chance of hurting, thwarting and spoiling human lives were seriously put out if, all the same, you behaved like a lady.[14]

Here is the language of disavowal, anticipating Mannoni's famous formulation. But beyond that, there is also something striking about Clarissa's sense of command and poise, how such bleak subject matter gets swept into view by way of a dependent clause, then dispatched with breezy defiance. This looks like the opposite of Mrs. Swithin's mental stagger through prehistoric swamps—Mrs. Dalloway, it seems, *has* found a form for these thoughts. And yet this form is also, in its own way, obviously insufficient. In its quippy, vaguely Wildean polish it might meet the needs of a social moment—which is how it is framed here, as a pose Peter remembers Clarissa striking—but at other moments, especially when Clarissa is alone with herself, we see how she struggles to maintain her poise:

> It rasped her, though, to have stirring about in her this brutal monster! to hear twigs cracking and feel hooves planted down in the depths of that leaf-encumbered forest, the soul; never to be content quite, or quite secure, for at any moment the brute would be stirring, this hatred, which, especially since her illness, had power to make her feel scraped, hurt in her spine; gave her physical pain, and made all pleasure in beauty, in friendship, in being well, in being loved and making her home delightful rock, quiver, and bend as if indeed there were a monster grubbing at the roots, as if the whole panoply of content were nothing but self love! this hatred!
> Nonsense, nonsense! she cried to herself.[15]

Vicki Tromanhauser connects this scene to Darwin, via Freud's theory of "organic repression" and anxieties about humanity's "animal constitution."[16] The decision to "behave like a lady" or calmly "decorate the dungeon with flowers" avails little when one feels oneself trapped inside a failing body. That phrase "nonsense, nonsense" is repeated a moment later, incantatorily, almost fetishistically, when she enters Miss Pym's flower shop, a place showcasing the easy commodification of nature and possessing its own hypnotically reassuring power:

> And as she began to go with Miss Pym from jar to jar, choosing, nonsense, nonsense, she said to herself, more and more gently, as if this beauty, this scent, this colour, and Miss Pym liking her, trusting her, were a wave which she let flow over her and surmount that hatred, that monster, surmount it all.[17]

It takes a moment before the meaning of this sentence becomes clear. That lag in comprehension occurs partly because of the lack of quotation marks and partly because of the grammatical ambiguity of the word "choosing," which can be used transitively (as it first appears) or intransitively (as the sentence actually demands). But just as the feeling of awkwardness or incongruity remains even after the pieces fall into place, the transitive direction of the verb lingers, because the question here, and throughout, is precisely *what* Mrs. Dalloway is choosing. She is in the act of choosing, and she is choosing *something*—choosing sound, color, consolation, reassurance, social connection. And she is also choosing, as the shadow sentence voices, *nonsense*; choosing to believe because it is what she wants to believe; choosing to behave like a lady *as if* it matters. The result is not so much belief in the meaning of floral decorations and proper comportment but the willing suspension of disbelief in such things. If, for the Victorians, the promise of capital-C Culture was that it could effect a "human separation from nature," as Kay Anderson puts it, for Woolf, the offer it extends is decidedly less grand: it promises to allow you to forget that its promises aren't real.[18] Circular, nonsensical, but perhaps no less compelling for all that.

Such are the experiences of Mrs. Swithin and Mrs. Dalloway reading Darwin and company, but these characters could be anybody. In fact, just after Clarissa leaves the flower shop, Woolf mentions a random passerby, "Mr. Bentley," who happens to be thinking about "Mendelian theory," among other things.[19] What Woolf is interested in is the way that, despite his iconic status, despite the fact that the "Darwin Industry" had by this time opened up major field offices in almost every possible area of inquiry, Darwin's work had somehow never really been fully reckoned with or integrated by the culture at large.[20]

An argument that I've been making implicitly thus far is maybe worth putting more squarely on the table now, which is that Woolf's fictional method uniquely lends itself to the representation, and even in some ways the *experience*, of denial. The sentence-level experiments with form I've been discussing make up part of a larger project in which a vast amount of narrative space is granted to registering the very smallest motions of the mind so that Woolf's texts are unusually alive to what we might call the phenomenology of attention. The granular, moment-by-moment representation of characters' experiences over the course of a single day (as is the situation in *Mrs. Dalloway*, *Between the Acts*, and the opening and closing sections of *To the Lighthouse*) means that she closely tracks an intricately, ceaselessly unfolding internal choreography of ideas, memories, moods, feelings, perceptions, and sensations, which necessarily includes both the emergence of random or unwanted thoughts *and* the various ways in which such thoughts get pushed aside, forgotten, rationalized, or otherwise evaded. And such narrative openness to experiential flux means that the dynamics of denial also structures the reading experience. A few pages after the scene at the florist's, Mrs. Dempster, a random passerby in Mrs. Dalloway's general ambit, notices an airplane flying overhead, and this trips off a series of thoughts and images:

> Away and away it went, fast and fading, away and away the aeroplane shot; soaring over Greenwich and all the masts; over the little island of grey churches, St. Paul's and the rest till, on either side of London, fields spread out and dark brown woods where adventurous thrushes hopping boldly, glancing quickly, snatched the snail and tapped him on a stone, once, twice, thrice.[21]

The sight of an airplane; the vision of London from an aerial perspective; the even wider shot of the surrounding countryside; then the sudden zoom down into the woods; then down even further to a single bird; then finally all the way to the small precisely numbered taps of its attack: chains of association can take someone anywhere at a moment's notice, and for reasons that are perhaps entirely inexplicable. It is, crucially, the non-human that is found at the end of this chain, the backgrounded reality that can be reached from anywhere because it is everywhere, emerging at once into the foreground of conscious experience. Woolf's use of free indirect discourse means it's not at all clear where, or even whether, Mrs. Dempster's voice stops and the narrator's takes over, but that very ambiguity suggests something pervasive, ambient, collectively shared about such trains of thoughts and imagined scenes. The non-human is there and not there; it moves in and through and among the minds of passersby, grading out, perhaps, into something more atmospherically available, as it were. We might put this in a slightly more tendentious way and ask who *owns* these thoughts. The answer seems to be everyone and no one; they are, in some ways, *dis*owned. This is true, on yet another level, of the motion of the narrative itself. Because we might ask further: Who is Mrs. Dempster anyway? Why is this person in the novel, and what, besides crossing her path, does she have to do with Mrs. Dalloway? Or what's the connection between the sharp vision of this snail-killing bird and the central events of the text—Mrs. Dalloway's preparations for the party, the return of her ex-lover from India, her strained relationship with her daughter, and so on? Does it matter? Is it a symbol for something? The novel will not say—it moves on without mentioning any of this again. The moment is vivid and precise and memorable; all the same, it does not connect in any obvious way to anything, either metaphorically or metonymically, so it is also there for the purposes of being forgotten.

One additional example, this time from her early novel *Jacob's Room* (1922), will help make these dynamics clear. At the end of the opening chapter, in a scene that resembles the end of the first section of *To the Lighthouse* (1927), a mother, Mrs. Flanders, finishes putting her children to bed and thus, it would seem, brings the events of the day to a close. But for a few moments, the narratorial eye keeps watching past the boundaries of the human story that would ordinarily organize its focus:

> There was a click in the front sitting room. Mr. Pearce had extinguished the lamp. The garden went out. It was but a dark patch. Every inch was rained upon. Every blade of grass was bent by rain. Eyelids would have been fastened down by the rain. Lying on one's back one would have seen nothing but muddle and confusion, clouds turning and turning, and something yellow-tinted and sulphurous in the darkness.[22]

This is followed a few paragraphs later by the chapter-concluding image of a crab trying and failing to climb the side of Jacob's toy bucket, which has been forgotten (by him and by us): "The child's bucket was half-full of rainwater; and the opal-shelled crab slowly circled round the bottom, trying with its weakly legs to climb the steep side; trying again and falling back, and trying again and again."[23] As occurs in *On Being Ill*, a different and potentially significant kind of awareness becomes available when the ordinary waking world is left behind; as occurs in *Mrs. Dalloway*, the extended glimpse of the non-human makes palpable the existence of a usually neglected ongoing reality that sits at a conspicuously awkward angle to the main story. It is in the breach rather than the observance of the usual procedures of narrative realism that we are made aware of what those procedures usually exclude, but without a clear sense of how, exactly, we are meant to include them.[24] In a diary entry from 1920, Woolf described the new conception for narrative form that would produce *Jacob's Room* two years later: "Having this afternoon arrived at some idea of a new form for a new novel. Suppose one thing should open out of another . . . only not for 10 pages but for 200 or so—doesn't that give the looseness & lightness I want; doesn't that get closer & yet keep form & speed, & enclose everything, everything?"[25] Such a ramifying, interconnective, total but (seemingly) non-hierarchical vision for the novel might well be called "ecological," distantly echoing Darwin's own representation of nature as a system of endlessly ramifying "relations" that can be traced "onwards in ever-increasing circles of complexity."[26]

We might mark a difference between this vision of connectivity and that of Henry James, as he describes it in his Preface to *Roderick Hudson*

> Really, universally, relations stop nowhere, and the exquisite problem of the artist is eternally but to draw, by a geometry of his own, the circle within which they shall happily *appear* to do so. He is in the perpetual predicament that the continuity of things is the whole matter, for him, of comedy and tragedy; that this continuity is never, by the space of an instant or an inch, broken, and that, to do anything at all, he has at once intensely to consult and intensely to ignore it.[27]

James wrote this preface in 1910, just ten years before Woolf's diary entry, but the gap (generational and otherwise) between the two writers is clear. For James, denial is a deliberate and necessary part of artistic practice: the novelist has to willfully ignore those aspects of reality that might lend a fuller or more complete vision of the world but that don't suit the purposes of his story. Realism depends upon a principle of exclusion or suppression—knowing it yourself ("intensely consulting") but, at the same time, not knowing it on the page ("intensely ignoring"). James does not explicitly frame this as about keeping the non-human excluded from the human story, but one need only have a passing familiarity with his novels to know that that is indeed one of the lines of demarcation that helps organize them. The vaguely paradoxical idea of "a geometry of his own" suggests the way in which the artist creates a factitious measurement system that is at odds with what is "really, universally"—and one might add "scientifically"—true but that is nevertheless made to *feel* true. Engaged in a "dire process of selection and comparison, of surrender and sacrifice," as he puts it, the novelist bears the entire burden of this denial himself.[28] Woolf, in contrast, puts that "dire process"—or much more of it—into the frame of the narrative; she shows the exclusions, the cuts, the broken places, and makes the reader confront the work of denial that gets taken for granted because it ordinarily goes on behind the scenes. And it is precisely this question of the work behind the scenes that, as we'll see in *To the Lighthouse*, aligns her development of an "ecological form" with her critique of English patriarchy and empire.

THE HOUSE DENIAL BUILT

The dynamics of denial are built into the very setting of *To the Lighthouse* (1927): a beach house overlooking the ferocious North Atlantic, which appears uneasily, alternately, and sometimes simultaneously, as both fraught central subject and disregarded backdrop. The sea—vast, inhuman, chaotic, overwhelming—is always there to be remembered, so it is also always there to be forgotten. Though a vacation home like the Ramsays' might be imagined as a retreat or an "escape," the irony is that its very location winds up trapping characters with uncomfortable realities they would perhaps sooner forget. The non-human thus makes itself felt in the narrative not through trains of associations running through the minds of characters (as in *Mrs. Dalloway*) or through sharp-angled narratorial interventions (as in *Jacob's Room*) but merely through its vast and unavoidable *presence* in the diegetic world of the narrative. The sea and its "meaningless plungings of water and wind," in

Wallace Stevens's phrase, functions as a kind of standing challenge to the significance of the human stories that would otherwise have our full attention.[29]

This dynamic is signaled most clearly in the opening pages of the novel, when Mrs. Ramsay realizes her thoughts have unexpectedly returned to the ceaseless sound of the ocean:

> Like a ghostly roll of drums [it] remorselessly beat the measure of life, made one think of the destruction of the island and its engulfment in the sea, and warned her whose day had slipped past in one quick doing after another that it was all ephemeral as a rainbow—this sound which had been obscured and concealed under the other sounds suddenly thundered hollow in her ears and made her look up with an impulse of terror. They had ceased to talk; that was the explanation.[30]

The novel thus establishes from the beginning a tension between the human concerns of the family and its guests and the non-human "background" that, with just the slightest shift in attention, might come flooding into the foreground. The barrier dividing foreground and background is permeable, inadequate, beset, and troped in the very physical structure of the house itself, with its drafts and creaks, its peeling wallpaper, and its open doors and windows that admit the elements. "Admit" indeed, in both senses: the house is, in a sense, a figure for language itself and how it is used to demarcate and provide form and clear a separate space for the human, while making it impossible to forget that it cannot actually do any of these things. What stands between Mrs. Ramsay and the uncomfortable thoughts that would wash into her mind is the mere sound of talk, the noises and gestures and motions of culture that can only temporarily muffle, can't seem to substantively frame or address, the non-human sounds coming from outside. The crucial point here is that it doesn't matter what is being said so much as *that* it is being said: that the "other sounds" function to cover up "this sound." Such a description, free-indirectly rendered, suggests that, on some level, Mrs. Ramsay is aware of her own processes of denial and recognizes how she deliberately takes form *as if* it is content. Mr. Ramsay and his interlocutors are busy discussing math, politics, art, whatever—the substance, she on some level understands, is irrelevant. Or rather, the content matters to Mrs. Ramsay only insofar as the men believe that it matters so that she can believe it matters, even though she knows it doesn't. Woolf sets all of this in motion so that we don't just see Mrs. Ramsay fooling herself but see her *recognizing* that she does this and, the next minute, allows herself to be fooled anyway. "What did it all mean?" she asks of the men's talk at dinner, then continues:

> To this day she had no notion. . . . She let it uphold and sustain her, this admirable fabric of the masculine intelligence, which ran up and down, crossed this way and that, like iron girders spanning the swaying fabric, upholding the world, so that she could trust herself to it utterly, even shut her eyes, or flicker them for a moment, as a child staring up from its pillow winks at the myriad layers of the leaves of a tree. Then she woke up. It was still being fabricated.[31]

Mrs. Ramsay falls for her own card trick, recognizing the emptiness of this sort of "intelligence," and, by extension, the entire world of men, but still allowing herself to trust it *as if* this world-spanning fabric is as real as the brown stocking she knits out of a sense of charity for the lighthouse keeper and his son. The contranym "fabricated" encodes a disavowal that mirrors the self-sustaining and self-canceling work of men, who are both making and "making up" their world. Just as we saw in *On Being Ill*, her small but tangible work of generosity gets denigrated in favor of unreal but extravagant masculinist claims.

The secret, of course, is that it is Mrs. Ramsay and those like her who actually "make" the world, who make all of this possible. And they do so, paradoxically, by concealing—even to themselves—the fact that they do. This is a realization Rachel Vinrace, the protagonist of Woolf's first novel, *The Voyage Out*, also comes to over the course of that novel:

> [Her father] was a great dim force in the house, by means of which they held on to the great world which is represented every morning in the *Times*. But the real life of the house was something quite different from this. It went on independently of Mr. Vinrace, and tended to hide itself from him. . . . It was her aunts who influenced her really; her aunts who built up the fine, closely woven substance of their life at home. They were less splendid but more natural than her father was.[32]

It is the women of the house who are in contact with "real life," who tend to the small but necessary material functions of human existence—cooking, washing, dusting, nursing, visiting "charwomen with bad legs"—and thus, in some basic sense, make everything else possible.[33] But that making is the very thing that allows Rachel's father to take as reality the illusion that that's what *he* is doing. It's only because he has so little to do with reality that he can imagine himself to be its architect. But where Rachel finds in this realization a new freedom to reimagine the categories through which she sees the world, to rethink the relation between foreground and background, Mrs. Ramsay finds in the dimly perceived threat of this realization only more reason to maintain the given hierarchical arrangements. The more arbitrary those arrangements come to seem, the more necessary they feel to her—the basic self-motivating paradox of denialism. While Mr. Ramsay wanders the beach believing he is confronting raw reality, pondering "the sea eating the ground we stand on" and other such things, he does so in order to picture himself a daring explorer, martyred on the peaks of knowledge: "the leader of that forlorn party which after all has climbed high enough to see the waste of years and the perishing of stars. . . . When the search party comes they will find him dead at his post, the fine figure of a soldier."[34] The masculine intelligence that "sustains" and "upholds" the world is actually all need and desire and dependency—is, in fact, being secretly sourced from elsewhere.

The shape of Mr. Ramsay's daydream is as Victorian as an antimacassar: there are hints of Tennyson's "Ulysses" as well as any number of frontier-braving romances, quest narratives, and scientific memoirs from the period. But also quintessentially Victorian, if less obviously so, is its dependence on what Zach Horton calls "scalar collapse"—an epistemological error he identifies in the work of Charles Lyell and other nineteenth-century scientists, in which the bridging of disparate scales is done in such a way as to "elide the differences between them."[35] By imagining himself equipped to witness something so utterly beyond the bounds of human sensory capacity and temporality as "the perishing of stars," Mr. Ramsay reimposes a human scale upon the cosmic span of Deep Time, when, clearly enough, the *loss* of a human scale is precisely what is most difficult to countenance. The passage echoes Huxley's similarly mythologizing vision of humankind's status in *Evidence as to Man's Place in Nature* (1863): "He has slowly accumulated and organized the experience which is almost wholly lost with the cessation of every individual life in other animals; so that he now stands raised upon it as on a mountain top, far above the level of his humble fellows, and transfigured from his grosser nature by reflecting, here and there, a ray from the infinite source of truth."[36] Where Mr. Ramsay fancies himself alone on the mountain, Huxley imagines the abstract figure of "Man" swollen to cosmic stature by means of the gathered attainments of civilization; both implicitly reestablish a sense of commensurability between the human world and that of the larger cosmos through what Timothy Clark calls "scale framing": "discursive practices that construct the scale at which a problem is experienced as a mode of predetermining the way in which it is conceived. Crucially here, to frame the scale at which one considers a problem is also sometimes a way of evading it."[37] Thus, the relationship between the scalar differences at issue and the imagery of "scaling" a mountain is more than a mere pun: the latter image, and all its associations with the Romantic sublime, is meant to feel like more of a change in perspective and proportion than actually is the case. The belief one is boldly and heroically confronting the "real" becomes the means of invisibly readjusting the scale to suit a preferred self-conception. It is the denial of denial, the insidious circularity of which structure of feeling plays no small part in the paralysis and exhaustion of the Ramsay household.

In short, Mr. Ramsay's sense of independence is entirely dependent on Mrs. Ramsay: this is the paradox that must be maintained but cannot be acknowledged, even though to maintain it, on some level, *is* to acknowledge it. Because Mr. Ramsay's insatiable need for sympathy and reassurance belies his self-construction, the very process of reassuring him both addresses and exacerbates the problem for which he needs the reassurance in the first place. The truth is known and not known, revealed and concealed in the same gesture, as denial is symbiotically coproduced between the two of them. But the energy required by that production process is unevenly distributed because the actual truth of the situation is differentially experienced. Mrs. Ramsay, "flashing her needles," knits together the fabric her husband is always unknowingly unraveling: "She assured him, beyond a shadow of a doubt . . . (as a nurse carrying a light across a dark room assures a fractious child), that it was real."[38] But where he is sustained by the imagined world she composes and recomposes, the story she is telling, her role as creator and sustainer physically taxes her in a way that cannot be falsified:

> Filled with her words like a child who drops off satisfied, he said, at last, looking at her with humble gratitude, restored, renewed, that he would take a turn; he would watch the children playing cricket. He went. Immediately, Mrs. Ramsey seemed to fold herself together, one petal closed in another, and the whole fabric fell in exhaustion upon itself, so that she had only strength enough to move her finger, in exquisite abandonment to exhaustion, across the page of Grimm's fairy story, while there throbbed through her, like the pulse in a spring which has expanded to its full width and now gently ceases to beat, the rapture of successful creation.[39]

Mr. Ramsay's entire sense of self, though it seems to dwell in the bloodless realms of metaphysical speculation, reputation management, and alphabetically charted progress, is actually sourced from Mrs. Ramsay's body. This is something that he literally does not see, and imaginatively cannot recognize, but that she of course cannot *not* recognize. As in *On Being Ill*, the body's limits and demands form a hard nub of reality that resists falsification; Mrs. Ramsay's body tells of her husband's infantile needs, his state of arrested development, his parasitical emotional life, though, crucially, never in so many words. As we have seen, Woolf's use of free indirect discourse puts us in a realm where the image of Mr. Ramsay as a nursing baby "drop[ping] away satisfied" resides in the nebulous zone where thoughts are shared, but not "owned" (to borrow Wood's term again), by narrator and character. The "he said" that follows the word "satisfied" suggests momentarily, but incorrectly, that he openly acknowledges his infantile need for sustenance from her; as the sentence continues, of course, the "he said" clearly applies to what follows ("that he would take a turn"). But similar to the "nonsense, nonsense" passage in *Mrs. Dalloway*, the fleeting grammatical uncertainty and the sudden shift from description of a concluded event to announcement of prospective action suggest that the truth of the situation is somehow ambiently, ambiguously available to him. For her part, Mrs. Ramsay also knows and does not know this truth: feels in her body what she cannot allow herself to say. Woolf makes this explicit on the next page, when she reflects on the fatigue Mrs. Ramsay experiences after the encounter with her husband:

Not that, as she read aloud the story of the Fisherman's Wife, she knew precisely what it came from; nor did she let herself put into words her dissatisfaction when she realised, at the turn of the page when she stopped and heard dully, ominously, a wave fall, how it came from this: she did not like, even for a second, to feel finer than her husband; and further, could not bear not being entirely sure, when she spoke to him, of the truth of what she said. Universities and people wanting him, lectures and books and their being of the highest importance—all that she did not doubt for a moment; but it was their relation, and his coming to her like that, openly, so that any one could see, that discomposed her; for then people said he depended on her, when they must know that of the two he was infinitely the more important, and what she gave the world, in comparison with what he gave, negligible.[40]

This passage is remarkable for the way it very precisely calibrates the ambiguous state of Mrs. Ramsay's understanding—syntactically hesitant, filled with litotic constructions ("she did not doubt") and verbs of cognition and meta-cognition in tension-laden configurations ("realizing," but not "knowing precisely;" refusing to put "into words" a feeling that is immediately labeled "dissatisfaction"; a sense of superiority, "feeling finer," that also makes her feel worse). Just as we saw with the term "fabricated," words and phrases are cleaved in auto-oppositional formations: in this case, "must know" signifies at once that which is self-evident—they "must know" his superiority because it is obviously true—*and* that which needs to be insisted upon or displayed because it's *not* self-evident—they "must know" in the sense of being *made* to know, because it is socially and psychologically imperative to maintain appearances. Such contrary energies in the language point to the fissures in Mrs. Ramsay's psyche, how she half-acknowledges her own awareness of the way she hides things from herself. This arrangement sometimes pays off in moments of pleasing forgetfulness, as when she can "shut her eyes" and trust the masculine intelligence "upholding the world" or when she has the feeling, a few pages later, of hovering "like a hawk suspended."[41] The imagery and language suggest ungroundedness, of course, a feeling of being loosed from the bonds of the earth so that maybe it doesn't matter that "the land is dwindling away" and getting swallowed up by the sea. But such rhetoric also suggests the "suspension" of disbelief, how Mrs. Ramsay, the authoress of the dinner party, can take as real her own elaborately constructed artifices. This includes, in a small strange twist, her seeming willingness to take credit for the meal she has very obviously not cooked herself: "How did she manage these things in the depths of the country?," William Bankes asks her after pronouncing the *Boeuf en Daube* "a triumph."[42] Though she doesn't linger on it, Woolf hints at an added dimension of denial, one rooted in class privilege, which underlines the way Mrs. Ramsay is positioned as both victim and beneficiary of this system.

There are, in a sense, two intertwined "plots" of denial that organize the dinner scene, both of which are set in motion and superintended by Mrs. Ramsay. The first has to do with the production of a shared sense of coherence and stability around the table, which depends upon her ability to manage conflict and the sharp edges of the various assembled personalities. But it also depends upon shared embodied experiences: the smells and taste of the main course, for example, or the mood-altering effect of lighting the candles, which turns the windows into mirrors: "Far from giving any accurate view of the outside world, [they] rippled it so strangely that here, inside the room, seemed to be order and dry land; there, outside, a reflection in which things wavered and vanished, waterily.... They were all conscious of making a party together in a hollow, on an island; had their common cause against that fluidity out there."[43] The Arnoldian symbolism couldn't be clearer—culture defines and organizes itself in opposition to the waves and flinging pebbles and seeming anarchy of nature—but the sense of opposition in Woolf's novel is utterly fantastical, a collective delusion made possible by the mirror of human self-regard. It is characteristic of *To the Lighthouse*—and a key, in my opinion, to its greatness—that it ironizes such feelings without discounting them; the production of denial is critically anatomized but with sympathy and an acknowledgment of just how difficult it is to know how to process unsettling and inhuman realities. This plot culminates in an epiphanic moment for Mrs. Ramsay, where she suddenly feels struck by the sense that "there is a coherence in things, a stability; something, she meant, is immune from change, and shines out (she glanced at the window with its ripple of reflected lights) in the face of the flowing, the fleeting, the spectral, like a ruby.... Of such moments, she thought, the thing is made that endures."[44] This is obviously an illusion, an indulgence, an experience of total self-enclosure disguised as a moment of breaking through to touch the "real"; all the same, Woolf doesn't seem to begrudge Mrs. Ramsay this feeling or fail to understand why she might feel the need to feel it.

The second plot has to do with how—or whether—this culture of denialism gets transmitted to the next generation, and it centers on the painter, Lily Briscoe. Lily adds yet one more layer of self-consciousness to the operations of denial: she plainly recognizes the oppressive gender dynamics at play around the dinner table and in the Ramsay household; she feels the expectation that she must assist in reproducing them; and she notices how she feels drawn to meeting that expectation in spite of herself. We see this most clearly in her relationship to Charles Tansley, whom she feels compelled, maybe even coerced, under Mrs. Ramsay's watchful eye, to placate and soothe, despite his open misogyny and gratuitous denigration of her and her work. The grotesquely uneven exchange between them replicates the Ramsay marriage: the man wounds yet considers himself the wounded party; the woman, meanwhile, is expected to play emotional nursemaid and pretend it never happened. Lily's self-awareness of the denialism at work is spelled out even more overtly in her reflections on Mrs. Ramsay's inveterate marriage plotting, which provides shape and focus and narrative structure to their days on the island:

> Such was the complexity of things. For what happened to her, especially staying with the Ramsays, was to be made to feel violently two opposite things at the same time; that's what you feel, was one; that's what I feel, was the other, and then they fought together in her mind, as now. It is so beautiful, so exciting, this love, that I tremble on the verge of it, and offer, quite out of my own habit, to look for a brooch on a beach; also it is the stupidest, the most barbaric of human passions. . . . This is not what we want; there is nothing more tedious, puerile, and inhumane than this; yet it is also beautiful and necessary.[45]

Like Mrs. Ramsay's reflections, Lily's thoughts are rendered in free indirect discourse but with the sonic balance clearly tipped toward the character rather than the narrator: the use of the first-person *I* rather than the usual free-indirect third-person *she* suggests we are verging on the direct representation of Lily's thoughts. One could easily imagine this passage, or parts of it, being set between quotation marks. In contrast to Mrs. Ramsay, whose awareness of the power dynamics she always keeps partly veiled from herself, Lily's thoughts are much more distinctly articulated and starkly diagnostic. We feel in her the gravitational pull of this culture as she struggles to attain escape velocity.

In short, what we find in *To the Lighthouse* is one of the most complex and precisely rendered articulations of denial in English literature. But I want to underline something that has been running implicitly through this reading, which is that Woolf consistently connects the denial of women's labor to the denial of non-human nature. The eco-theorist Val Plumwood's discussion of "denied dependency" reads almost like a gloss on Woolf's novel:

> One of the most common forms of denial of women and nature is what I will term backgrounding, their treatment as providing the background to a dominant, foreground sphere of recognised achievement or causation. This backgrounding of women and nature is deeply embedded in the rationality of the economic system and in the structures of contemporary society. What is involved in the backgrounding of nature is the denial of dependence on biospheric processes, and a view of humans as apart, outside of nature, which is treated as a limitless provider without needs of its own. Dominant western culture has systematically inferiorised, backgrounded and denied dependency on the whole sphere of reproduction and subsistence. This denial of dependency is a major factor in the perpetuation of the non-sustainable modes of using nature, which loom as such a threat to the future of western society.[46]

Plumwood, like Rachel Vinrace, like Lily Briscoe, like many other characters in Woolf's fiction, finds in the spatial metaphors of foreground and background a way of thinking about the wildly imbalanced distribution of attention, regard, and value. For Lily, of course, this is made plain in the way she works on her painting, finding in the freedom of the space of the canvas a refuge from the hierarchical arrangements of the dinner table, the house, the culture at large. And in Mrs. Ramsay, we clearly see someone "backgrounded" and treated as a "limitless provider without needs of her own"—as the narrator says after the work she does consoling Mr. Ramsay, "There was scarcely a shell of herself left for her to know herself by; all was so lavished and spent."[47] The connection between women and the non-human runs directly through the body, as we see the toll in material energy taken by the maintenance of the fictions Mr. Ramsay needs to feel better and all the various metaphors applied to her that indicate the non-human part of her being: she is a shell, a flower, a fruit tree, a spring, a "rain of energy, a column of spray."[48] The world of language has a material resource basis, so not only does Mrs. Ramsay's stream of reassuring storytelling come at a physical cost to her body, but also the work of denial *itself* appears literally as a form of labor. The crime, on some level, is the cover-up.

As Plumwood notes, a one-way system of resource extraction must inevitably exhaust itself, and this is indeed what happens to Mrs. Ramsay. The description of her death is of course famously rendered in a brief parenthetical in the next section "Time Passes," and though that moment has been much remarked upon in criticism of the novel, it is perhaps worth quoting again in full for the way it captures the shocking end point of denialism: "[Mr. Ramsay, stumbling along a passage one dark morning, stretched his arms out, but Mrs. Ramsay having died rather suddenly the night before, his arms, though stretched out, remained empty.]"[49] The cause of her death is left unspecified, suggesting that she dies not so much from some undisclosed ailment or disease but by an unsustainable system of relations that has finally wrung her dry. Indeed, the very arrangement of the sentence announces itself as a networked system of forces in which her death is formally embedded: the use of brackets and the subordination of Mrs. Ramsay in a grammatically dependent but semantically central clause makes shockingly plain the structuring tension between centrality and marginality, between a hierarchizing order and the actual production of meaning, that has defined their marriage all along. Mrs. Ramsay is an emptiness, a void that her husband's outstretching arms encircle syntactically as well as materially, because, as we have seen, there is something about the verbal demands of the relationship that has a cost in material life. In other words, the sentence is not merely the familiar instance of form matching content but a place where the interchange of verbal and material realities, of expressive acts and bodily resources—an interchange Woolf has been developing throughout the novel—reaches a crisis point. The end of falsifiability. The chiastic stretched out/ arms // arms/stretched out conveys replication and transformation, as Mr. Ramsay, like a man who mechanically takes another step upward after having already reached the top of the stairs, is startled awake by the feeling of finding nothing supporting him. The somnambulant routines that, in *On Being Ill* bespeak a kind of habitual denial of the body, are literalized in this scene, as he sleepwalks straight into reality.

"Time Passes" both brings denialism to its inevitable end point in the death of Mrs. Ramsay and dramatically widens the picture to suggest how it works in the culture more broadly. This expansion includes, for one thing, a much more thorough consideration of the "downstairs" life of the servants that only gets glancing treatment in the previous section. Like the porter in *MacBeth*, Mrs. McNab comes from nowhere to take center stage, and with her comes a much more direct and literal depiction of bodily toil and the physical cost of maintaining a sense of household order: "How long, she asked, creaking and groaning on her knees under the bed, dusting the boards, how long shall it endure? but hobbled to her feet again, pulled herself up . . . and began again the old amble and hobble, taking up mats, putting down china, looking sideways in the glass."[50] The insistence on the wear and tear of work is matched by, and gives expression to, a fuller consideration of the ferocious entropic action of the non-human world, now no longer muffled by the various insulations of class privilege. The forces of decay with which Mrs. McNab must contend are ceaseless and ubiquitous: the woodwork swells and cracks from the moisture; a bit of wallpaper flaps off the wall; the books rot on the shelves; the saucepan rusts; a floorboard on the landing suddenly springs out of place. Although the first section of the novel acknowledges these non-human forces and thus prepares the way for "Time Passes," there is something still shocking—I feel it no matter how many times I read the novel—about how relentlessly the house gets dismantled, how unsparing and ferocious Woolf's account of the change is. All the descriptions in the first section of the sea eating the land, the stars perishing in the sky, and the churning noises of the waves in the background now feel quaintly lyrical and soft-pedaled compared to this. What had been framed a back-and-forth battle between human structures of meaning and the non-human is now revealed as a rout, a massacre.

And, of course, it has always been thus. Like Mrs. Swithin's book on natural history, "Time Passes" reveals what is well established about the nature of the world; it delivers its shock via what we already know. Everything is headed inexorably in the same downward direction, yes; time works everywhere to weaken and corrode and split and loosen and topple—we know this; there is no quarter, no refuge, "no safety . . . no guide, no shelter," as she puts it: understood.[51] All of those things are subjects of the first part of the novel. But by suddenly thrusting the non-human upon our attention in this altogether new and bracing way, Woolf makes us feel how the conventions of ordinary narrative realism, the ones that have carried us through the book thus far, have quietly framed up the world in a human-scaled, tacitly reassuring way and thus might be seen as a means of evading the "real" under the guise of maintaining a scrupulous fidelity to it. We've watched with amusement and sympathy as poor Mr. Ramsay marched around believing himself to be in touch with bare reality. We've seen how that belief, the idea that he is free of denial, is in fact the very mechanism by which it has taken root in him. But now, in "Time Passes," the dynamic we observed becomes one we experience as our own privileged vantage point gets washed out from beneath our feet, as it were. As with so many things in this novel, we see the *work* that must be done—that has been happening all this time—to make the real seem real. For Mr. Ramsay, it's the work of Mrs. Ramsay; for Mrs. Ramsay, it's the work of the servants; for the reader, it's the work of the realist novel itself.

WITH THE LAMPS OUT, IT'S LESS DANGEROUS

But there is one way in which Plumwood's account of denied dependency might seem an imperfect fit with Woolf's novel, and that is its identification of the wider environmental consequences of the "backgrounding" of nature: "This denial of dependency is a major factor in the perpetuation of the non-sustainable modes of using nature, which loom as such a threat to the future of western society."[52] We've seen the threat denial poses to Mrs. Ramsay, and though we might read her as some kind of allegorical representation of a used-up world, the overwhelming emphasis in all three sections is not on the vulnerability of non-human nature but on its imposing, remorseless power and the steps everyone takes to manage their feelings about that. In some ways, this is perfectly in keeping with the novel's sense of a Victorian past-which-is-not-past, since, as we've seen, the things that so unsettled Tennyson, Arnold, Ruskin, and the rest—the vastness of geological ages, the undirected course of evolutionary history—still have yet to be fully reckoned with.

Although Woolf says nothing about pollution or deforestation or habitat destruction or any of the environmental consequences of industrial growth and accelerating fossil-fuel use, there are intimations of the Anthropocene in *To the Lighthouse*, some of which feel continuous with the environmental vision of the preceding era, some of which would seem to mark an important break, a new way of imagining the human situation on the planet. The rust and water and wind let loose in the pages of "Time Passes" are presented not merely as elemental forces or entropic processes—though they are clearly those things too—but as a strange and volatile mixture of the natural and the human: "And now in the heat of summer the wind sent its spies about the house again"; "the prying of the wind, and the soft nose of the clammy sea airs, rubbing, snuffling, iterating, and reiterating their questions."[53] At times, as in the second example, the personified escapades of the winds in "Time Passes" have an almost Dickensian feel to them. They seem to have more of the rowdy exuberance of what we find in *Martin Chuzzlewit*—"[The wind] slammed the front door against Mr. Pecksniff, who was at that moment entering.... The boisterous rover hurried away rejoicing, roaring over moor and meadow, hill and flat, until it got out to sea, where it met with other winds similarly disposed, and made a night of it"—than, say, the pagan majesty of Shelley's West Wind, or the mortal intimacy of Emily Brontë's Night Wind, or the punishing fury of Conrad's various ocean squalls and typhoons.[54] The weather in Woolf, as in Dickens, is not just human but all-too-human: meddlesome, headstrong, nosy, irrepressible. I'm not trying to make the case that "Time Passes" is funny, exactly, but rather that, as in Dickens, its depiction of the non-human has a metaphorical excess about it, a freakishness that, however semi-serious the tone, bespeaks deeper uncertainties about the categories that usually demarcate the human world from everything else. In both writers we find not so much the unnatural but the "abnatural," a term coined by Jesse Oak Taylor, which, in his words, "characterizes those moments in which nature appears other to itself, beside or outside itself... wherein everything from the bloodstream to the weather bears the traces of human action."[55] This unruly and evocative mixture of the human and the non-human is a hallmark of Dickens's style, of course, and part of what makes his novels, for many critics, such important early literary registrations of the onset of global environmental crisis. In *Our Mutual Friend*, for example, the wind and the productivist commercial spirit of London merge into a hybrid agent of chaos: "The grating wind sawed rather than blew; and as it sawed, the sawdust whirled about the sawpit. Every street was a sawpit, and there were no top-sawyers; every passenger was an under-sawyer, with the sawdust blinding him and choking him."[56] Neither entirely human nor entirely non-human, the city is becoming anti-human, a place where intentional local activity has mindless and dislocating aggregated results. "The wind sounds up here," one of the characters remarks, "as if we were keeping a lighthouse."[57]

Both Dickens and Woolf put the abnatural to work as part of larger critiques of capitalist modernity, but where the former's target is England's heedless commercial and industrial growth, conspicuously on display in the city, the latter's is a militarized imperial world order felt at the fringes of perception:

> But slumber and sleep though it might there came later in the summer ominous sounds like the measured blows of hammers dulled on felt, which, with their repeated shocks still further loosened the shawl and cracked the tea-cups. Now and again some glass tinkled in the cupboard as if a giant voice had shrieked so loud in its agony that tumblers stood inside a cupboard vibrated too. Then again silence fell; and then, night after night, and sometimes in plain mid-day when the roses were bright and light turned on the wall its shape clearly there seemed to drop into this silence, this indifference, this integrity, the thud of something falling.[58]

The force that loosens the shawls and cracks the tea-cups in the cabinet has been the protagonist of "Time Passes" up until this point, but now something has changed: the distant concussion of armaments coming from the North Atlantic, which lend their vibrations to the wearing action of wind and sea spray. Where the mingled sound of human voices once drowned out awareness of the waves for Mrs. Ramsay, they now accumulate on an entirely new scale and take on material form. The waves of water have become waves of sound, crashing through the walls and into the recesses of the house, washing through the fetishized appurtenances of the culture. The "giant voice" thus merges back with the non-human elements even as it expresses something of the outsized proportions of collective human action. This is immediately followed by another concussive blow delivered via another shocking parenthetical: "[A shell exploded. Twenty or thirty young men were blown up in France, among them Andrew Ramsay, whose death, mercifully, was instantaneous.]"⁵⁹ In retrospect, we see that the Great War has been quietly drawing nearer this whole time under the cover of Woolf's metaphors. Whether in the form of trees described as "tattered flags," or winds as "spies," or the stray sea airs as the "advance guards of great armies," the signs have been there, hiding in plain sight.⁶⁰ In many ways, the war is the novel's central subject, but it has remained concealed, backgrounded, until this moment; the explosion releases, as it were, a kind of building narrative tension and expresses another inevitable end of denial.

For Woolf, the First World War was an object of widespread denialism. Real, but not realized; known and not known, it is the catastrophe that is coming, the catastrophe that has come, the catastrophe that, even while it is happening, somehow cannot be integrated into a coherent picture of everyday life. This is a structure of feeling Woolf noticed in herself as well as in the culture at large: in her diaries, she writes about how the war could at once seem to be of grave and overwhelming importance and not entirely real:

> We are just (4. P.M.) off to Lewes to get a paper. There were none at breakfast this morning, but the postman brought rumours that 2 of our warships were sunk—however, when we did get papers we found that peace still exists—save for a stop press message that England has joined in. It is rather like Napoleonic times I daresay, and being a Bank holiday of course makes us more remote from life than ever.⁶¹

In Karen Levenback's account of Woolf and the Great War, she comments that her response was characterized, especially in the early years of the conflict, by "a sense of disbelief, one that seemed offset with something like expectation, as she both received and sought out information about the war."⁶² Even after experiencing a German air raid in London, Woolf wonders at her own feeling that it still "seems utterly impossible that one should be hurt."⁶³ This is the perspective of privilege—of the noncombatant, of a member of the upper class, of an inhabitant of a country not crisscrossed with trenches and wire. And while Woolf recognizes the shaping force of that privilege, she finds herself unable to divest herself of it. And at times, it must be said, she even seems to deliberately retreat into it, responding to the crisis by a misanthropic rejection of the world and her own connection to it. In her diary, she describes a day spent reading Wordsworth in the garden at Asheham House, in Sussex: "The daffodils were out & the guns I suppose could be heard from the downs. Even to me, who have no immediate stake, & repudiate the importance of what is being done, there was an odd pallor in those particular days of sunshine."[64] Woolf tries to lower the stakes to a kind of mild aesthetic disturbance, the interruption of an idyll for a mind that would otherwise be wandering lonely as a cloud. There is a studied casualness in that "I suppose" and something more than a little willfully evasive in her "repudiation" of the importance of what was happening at this particular moment in the war. Woolf wrote this on April 8, 1918, a few weeks after the beginning of the German "Spring Offensive" on the Western Front and the day after the commencement of the Battle of the Lys in Flanders.

This response, what Levenback calls the "illusion of immunity," was on some level a by-product of the unavoidably mediated access to the war for those not directly participating. On another level, though, the illusion of immunity was not a by-product but a *product*, an effect deliberately manufactured and maintained by the culture. In a contemporary account of shopping in London during the war, an anonymous writer in the journal the *Academy* describes how the windows function to shift attention away from

> the death and desolation that reign in the cities on which battle has set its terrible seal. Each beautiful fabric displayed, or product of the lands across the ocean, has a new value of its own. It means that looms are working and factory chimneys smoking; women and children warmed and fed and housed amid the tribulations of war. . . . Within were beauty and order and steady business, nothing feverish in its atmosphere, but a pervading sense of well-being.[65]

This was, as Elizabeth Outka memorably puts it, "using the windows, in effect, to hide the war."[66] The key point here, and the crucial thing that "Time Passes" shows, is that a culture of denial is not simply a response to crisis; it is itself, on some level, the source: producing the problem it then cannot bear to look at. In her novel *Not So Quiet: Stepdaughters of War* (1930), Evadne Price (writing under the pseudonym Helen Zenna Smith) describes the process by which firsthand accounts of the realities of the war were met with a denialist skepticism by male audiences at home:

> You live in a world of cold sick fear, a dirty world of darkness and despair—that you want to crawl ignominiously home from these painful writhing things that once were men, these chattered, tortured faces that dumbly demand what it's all about in Christ's name—that you want to find somewhere where life is quiet and beautiful and lovely as it was before the world turned khaki and blood-coloured—that you want to creep into a refuge where there is love instead of hate. . . . Tell them these things; and they will reply on pale mauve deckle-edged paper calling you a silly hysterical little girl—"You always were inclined to exaggerate, darling."[67]

Price's novel was originally commissioned by the London publisher Albert E. Marriott as a parody of Erich Remarque's *All Quiet on the Western Front* (1929)—he asked her to write a lightly comic send-up of women's experience of war under the proposed title "All Quaint on the Western Front." But we can see both in the previous passage and in the kind of book Price actually wrote her outraged repudiation of a male cultural establishment that would discount the work of women in the war effort (the protagonist of her novel is an ambulance driver) and brush aside the emotionally and psychologically destabilizing effects of that work. The man with the mauve stationary is, to borrow a contemporary term, "gaslighting" his correspondent, but Price's interest here, in playful tension with such comic specificity, is how this operates as a generalized cultural phenomenon—a widespread and even systematic effort to refuse, downplay, discount, in a word, deny—the reality of what was happening (in the diegetic world of the novel) and of what had just happened (in her own experience of the aftermath).

This is what Woolf means when, in a letter to Margaret Llewelyn Davies, she writes: "I become steadily more feminist, owing to the *Times*, which I read at breakfast and wonder how this preposterous masculine fiction keeps going a day longer."[68] The war is a "fiction" inasmuch as it doesn't quite seem real or believable and comes delivered in regular installments like some bad Victorian serial; but it is also a fiction in the sense that it *is* the reality actually being written for the world, the narrative men are collectively authoring and making everyone pretend makes sense. As with Mr. Ramsay and friends, it is a "fabrication." Their story seems to her narratively unsustainable—it seems as if it can't continue "a day longer" because it is being penned by hack writers who have lost control of the plot and who by now, surely, have exhausted their audience's ability to suspend disbelief in their powers.

The resonances between the war denialism Woolf describes and that of today's climate denialism aren't hard to see. In both cases, a crisis that is global in scope is differentially experienced according to one's location, class, nationality, and other forms of identity and position; insofar as its effects are accessible to those relatively insulated by privilege, it is often only through various technologies and forms of mediation—newspaper reports, wire services, statistical tables, anecdotes—which, at their worst, distort and conceal the situation. But even at their best, even when the mediated accounts are transmitting necessary and difficult truths, even when the reality thus presented might be fully acknowledged and, in some ways, even be understood as somehow more fundamentally "real" than the reassuring, quotidian day-to-day life in Bloomsbury, or in a vacation home in the Hebrides, or wherever, that reality nevertheless does not quite sink in or gain the kind of purchase on the imagination that would seem commensurate with the urgency and horror of the situation. For the rounds and routines of ordinary life, the material environment of shops and houses and cafés and roads, what Timothy Clark calls "denial in concrete" and what Woolf calls the experience of "business going on as usual" continue to have their reassuring, even narcotizing effect, which doesn't necessarily wear off just because you see how the game works.[69] The promises of order, progress, stability, control made in shop windows and bank lobbies and well-paved roads, not to mention in all the various fictions still being generated on behalf of global capital and its imperial designs, are being offered as if they still might work out, as if they have not been revealed as more than simply false but as the very source of the problem.

One could go on drawing comparisons between the structure of what I am calling "war denialism" in the early part of the twentieth century and that of climate denialism today; but of course, the relationship between these responses is more than just analogical; it's also historical. In *Facing Gaia*, Bruno Latour describes how the onset of the Great War both resembles our contemporary response to the climate crisis and is itself an event that reveals the compounding effect of denialism over the long durée:

> The alarms have sounded; they've been disconnected one after another. People have opened their eyes, they have seen, they have known, and they have forged straight ahead with their eyes shut tight! If we are astonished, reading Christopher Clark's *The Sleepwalkers*, to see Europe in 1914 hurtling towards the Great War with its eyes wide open, how can we not be astonished to learn retrospectively with what precise knowledge of the cause and effects Europeans (and all those that have followed the same path since) have rushed headlong into this other Great War about which we are learning, stunned, that it has already taken place—and that we have probably lost it?[70]

In Latour's account, the Great War both shows the way "alarms" get disconnected and was *itself* an unheeded alarm indicating the dangerously unsustainable foundations of the emergent capitalist world order. The contrary mix of metaphors he uses—people have "opened their eyes" but also have them "shut tight," and then have them "wide open" again as they forge and hurtle and rush headlong while somehow also sleepwalking—is also a feature of the book he cites: Clark's introductory account of his eponymous sleepwalkers is of a European officialdom that "walked toward danger in watchful, calculated steps."[71] Conscious and unconscious, eyes open and closed, asleep yet watchful—such oppositional descriptions are reflections of the strangely mixed condition of denial, of knowing and not knowing at once. They also reflect the complex and uncertain status of agency in the time of global crisis, in which the psychological apparatus of the individual human subject is of real but limited heuristic value in conceptualizing collective action and decision-making. Woolf's own well-known interest in representing various gradations of states of sleep and wakefulness, which we have already seen in the depiction of a bereft Mr. Ramsay, might take on a different kind of historical and even environmental significance if we see it as an attempt to make sense of a world being shaped by the twilit half-mindedness of aggregated human power.

As Timothy Morton argues, the Great War is so famously difficult to grasp because it demands a kind of ecological thinking about distributed agency and complexly networked sequences of cause and effect traveling across disparate scales and dimensions: "The more you know about the origins of the First World War, the more ambiguous your conclusions become. You find yourself unable to point to a single independent event."[72] Meanwhile, as the environmental historian R. S. Deese argues, the global scale of the war makes it a key inflection point in the dawning awareness of what would later be termed the Anthropocene: "It was the advent of total war between sovereign nation states, with all of its attendant destruction and technological innovation, which convinced many that our species had now attained the power to transform nature in fundamental ways."[73] As we have seen, the previous century was actually pretty convinced of this too, but its sense of this newfound human power too often came packaged in ways that seem remarkably naïve and dangerous. Recall William Kingdon Clifford's remark that "the creature of clay which you despise is the Lord of Nature and the Measure of all things": the paradoxical rhetoric of this remark at once conceals and exposes the illogic of this kind of tendentious scale framing.[74] A sense of cosmic insignificance leveraged into a dream of control; a belief in human animality made into the basis of a telos of transcendence. Darwin delivered a decisive blow to the idea of human exceptionalism at the same time that he seemed to show a way of dodging it. Or, we might say, *by* delivering such a blow, he made it feel like it needed to be dodged; made it feel like it still could be, even after it had landed. The discourse of evolution, pressed into service by capital and empire, underpins what Heidi Scott terms the discourse of the "Good Anthropocene": a Promethean belief that humanity was on the threshold of achieving an entirely new kind of directive power over nature.[75] For Woolf, as for so many, the Great War was the event that forced this situation into crisis, that revealed the contradictions and incoherencies that were always there to be seen but had been suppressed, sidestepped, evaded, rationalized, and denied. The conflict that convulsed the globe came into being not so much through any decisive act of will or cogent strategy but through a series of uncoordinated decisions and chains of unintended consequences that fanned out through an intricately networked global political order.

This question of global agency and human decision-making that the war made newly visible through the terrifying new category of "total war" is made a central problematic of *To the Lighthouse* and its exploration of the multileveled denialism that defines the Ramsay household. And that problematic, I would argue, is what makes the novel something of a hinge—a point of both continuity and discontinuity—between all the various evasions and rationalizations that define the response to Darwinian ecology and those that define the response to the climate crisis. In the transition from the "The Window" to "Time Passes" hangs a question that is no less pressing because of how strange and entirely unanswerable it is: What is the relationship between the denialism of the first section and the onset of the war? Or, what responsibility does the Ramsay family bear for what happened to the world? An absurd question on some level, with an easy answer: the Ramsays obviously did not cause World War I. And yet, on another level, considered through the vexed and uncertain status of agency in this era of mass-produced climatological unraveling, we must wonder—Woolf asks us to wonder—what role this culture of denialism, which the elder Ramsays embodied, defended, and sought to transmit to the next generation, played in bringing into being the conditions that made the war possible. Any individual person's everyday activities both do and do not matter, materially, to the climate crisis—they are both without any appreciable effect and also (all the same) a form of participation in the aggregate that *is* the problem. It is, in some ways, the other side of the knowing and not-knowing split of disavowal that we have been tracking: individuals are responsible and not responsible, just as the Ramsays did and did not cause the war. Crucial to this point is Woolf's fictional method, which at once offers and withdraws its allegorical significance. That is, Mr. Ramsay is clearly Victorian Sage, the Patriarch, the Tyrant, the Oedipal Father; he is the archetype and embodiment of a whole cultural order, and the disputes over the trip to the lighthouse bespeak a much wider struggle over the phallus, over authority, masculinity, meaning, teleology, and all kinds of other things. Such allegorical resonances—one could go on for pages detailing them—are clearly there, and they have, in a sense, a force-multiplying effect insofar as they suggest that the dynamics at work in the Ramsay household can stand in, on some abstract level, for what goes on in innumerable households across the country. At the same time, Mr. Ramsay is also just one specific, sad, old man, and the lighthouse is just a building with a light at the top. This hermeneutic uncertainty is, to borrow from the passage from Morton quoted at the beginning, part of Woolf's "radical openness" that involves confronting a world that poses urgent questions of perspective and meaning making that can have no final answer. That openness, I also submit, defines one of Woolf's signal achievements in this novel, which is the mix of caustic criticism and generous sympathy with which she handles the all-too-human problem of denial.

CONCLUSION
FULL DISCLOSURE

THIS BOOK OWES A PROFOUND debt to an erstwhile Victorianist whose extraordinary series of books beginning in the early aughts vividly connects the soft denial of evolution with the environmental crises of the present but whose name has not yet appeared in its pages. Margaret Atwood's *MaddAddam* trilogy is, among other things, a deeply imagined investigation of the pernicious effects of a culture of denial in which, as one of her characters (free indirectly) puts it, "Everybody knew. Nobody admitted to knowing. If other people began to discuss it, you tuned them out, because what they were saying was both so obvious and so unthinkable. *We're using up the Earth. It's almost gone.*"[1]

Atwood's trilogy is at the end of this book, but it was at the imaginative origins of this project. These are novels I picked up, closed in horror, picked up again, reread, taught (not necessarily in that order), scribbled notes in, spilled coffee on, and bought fresh copies of, for a long time before I began writing in earnest. Their importance for me resides in the way they lay bare the deeply contradictory place of Darwin in the Western imaginary and the self-consuming dynamic by which a culture of denial reinforces itself by bringing into being the very things it purports to defend against. Atwood is also a novelist with deep ties, imaginatively, aesthetically, thematically, to the nineteenth-century past. I called her a "Victorianist" mostly for the cheap rhetorical effect but also because she pursued a literature PhD at Harvard, working on a (never-completed) dissertation called "Nature and Power in the English Metaphysical Romance of the Nineteenth and Twentieth Centuries," which discussed H. Rider Haggard and George MacDonald, among others.[2] In essays and reviews, she has signaled her abiding interest in Mary Shelley, Robert Louis Stevenson, H. G. Wells, and any number of other nineteenth-century writers and has frequently remarked on her early education steeped in the Victorian novel as well as a childhood obsession with *Wuthering Heights*.[3] Atwood also connects her own political views to their roots in nineteenth-century Romantic anticapitalism: "I'm a Red Tory. To get a fix on this category, you have to go back to the 19th century. The Tories were the ones who believed that those in power had a responsibility to the community, that money should not be the measure of all things."[4] On a formal level, we can clearly see in *Oryx and Crake* the ways in which she plays upon the conventions of the nineteenth-century bildungsroman by reversing their usual procedures: the story of socio-personal integration becomes one of an asocial, bodily disintegration; the journey from countryside to city is flipped into a flight away from the collapsing structures of the urban world back into the wilderness; the protagonist's maturation process, which conventionally allegorizes the historical development of the nation, is here used to track a parallel sinking downward into darkness, loss, cultural forgetting, and social dissolution. I could go on. The point is that these paradigmatic tropes, deployed by Dickens, Charlotte Brontë, George Eliot, and any number of other nineteenth-century novelists, are all reconstituted through the act of upending them. I might also add that her novels strike the kind of balance we often associate with the great Victorian novels, where popularly accessible, plot-driven narrative meets serious social diagnosis and theorization. Atwood has written about the "profound shock to the Victorian system" delivered by *The Origin of Species* and *The Descent of Man*, how it disrupted the "kindly Wordsworthian version of Mother Nature."[5] Wordsworth and Darwin serve as familiar shorthand for the before-and-after pictures of nature; clearly, in her hands, the novel form remains a way to register the continuing aftershocks of the shift from one to the other.

Dipesh Chakrabarty writes that the challenge for his fellow historians at this moment is that "the crisis of climate change calls for thinking simultaneously on both registers, to mix together the immiscible chronologies of capital and species history."⁶ Atwood's trilogy can be seen as an attempt to work though the representational challenge posed by these disparate narrative frameworks for understanding the human—their simultaneous mixture and separation—in the form of a speculative future history. In a sense, the books project forward to the point where these twin histories are accelerating toward their own terminal convergence, where the meeting place of capital and species becomes unmistakable because the two become fused in totalizing and irreversible ways. We can see it happening, for example, in the many profit-driven innovations in bioengineering and genetic manipulation that, as Atwood takes pains to show, are decisively altering the evolutionary trajectory of both humans and any number of non-human animal species and re-ordering all aspects of life on the globe. And yet, the novels are also about the stubborn *feeling* of "immiscibility," about all the ways in which her characters both intensely know, and intensely refuse to know, how the logic of capitalism has turned the natural history of the species into a mass suicide event.

This is especially the case in the first book, *Oryx and Crake*, where both the obviousness of the threat of environmental collapse and the methods employed to tune it out have been pushed to mutually reinforcing extremes. One of Atwood's chief targets here is what Adorno and Horkheimer famously dubbed "the culture industry"—the various forms of popular entertainment whose main product is a passive and homogenized public of consumers. As Adorno and Horkheimer put it, "Those in charge no longer take much trouble to conceal the structure, the power of which increases the more bluntly its existence is admitted."⁷ In *Oryx and Crake*, mass entertainment doesn't attempt to conceal the warning signs of environmental breakdown or impending social collapse; on the contrary, it emphasizes such things in order to use them as fodder for video games, TV news, and the like: "He might go to a movie at the mall, just to convince himself he was part of a group of other people. Or he'd watch the news: more plagues, more famines, more floods, more insect or small-mammal outbreaks, more droughts, more chickenshit boy-soldier wars in distant countries."⁸ All experience is flattened out into a uniform, depthless field of ironized entertainment, treated as simultaneously real and unreal, known and not, unquestionable and unthinkable. In the pre-apocalyptic sections of the novel, culture operates according to the denial-by-admission logic of disavowal, creating one, giant, mediating defensive shield as it also, symbiotically, activates the defense mechanisms of its audience members to keep them frozen in place. After the collapse occurs, the protagonist, Snowman (aka Jimmy), reflects on this situation: "There had been something willed about it though, his ignorance. Or not willed, exactly: structured. He'd grown up in walled spaces, and then he had become one. He had shut things out."[9]

This passage contains, in a nutshell, so much of what *Climate of Denial* has been about. Atwood here shows the disavowing subject reflecting on his own process of disavowal, adding yet another layer of self-awareness that changes nothing. Some of the disturbing power of the novel inheres in the way this self-awareness can so freely survey itself and its history while it helplessly circles the drain. Motion without movement, to borrow a phrase from the Introduction. But one of the reasons for Snowman's incapacitation in this moment is that he is trying to define for himself what is a necessarily undefinable situation: the place where his own sense of agency blurs into the operations of a larger system. To what extent was his denialism an expression of his individual will, and to what extent was it simply a product of the structural forces shaping his mind and his life, making him believe he ever had a will in the first place? Where is the line, is there a line, and how could he even know? Note the use of our old friend free indirect discourse, how it allows Atwood to open a strategic ambiguity in the word "structured," which can, according to the grammar of the sentence, be read as either active or passive. "Structured" is, on the one hand, rhetorically paralleled with the active (grammatically and semantically) "willed," suggesting that it is Snowman himself who did the structuring, just as it was he who did the willing. On the other hand, the oppositional "not" construction prepares the way for a fuller kind of reversal, where "structured" might be imagined as something done *to*, not *by*, him and thus the exact opposite of "willed." If a person is just a product of the culture, there is no longer any distinction between deception and self-deception, since the notion of "self" is itself already part of the deception.

This is almost too obvious to point out, but just as there is no resolving this syntactical ambiguity, there is no resolving the existential ambiguity for Snowman about whether—or even the extent to which—he has been an agent in the events of his own story. He is asking himself a version of the question Dickens's David Copperfield famously poses at the beginning of his narrative—"Whether I shall turn out to be the hero of my own life, or whether that station will be held by anybody else, these pages must show"— except that he keeps asking it all the way up through the final paragraph, knowing full well that no pages will be written and no answer will show itself.[10] The deferral is the point: the ambiguity in a word like "structured" keeps such questions open and thus keeps the internal motion going as the external world falls apart. In *David Copperfield*, the action of deferred unveiling is inaugurated in the opening sentence, as it comfortably posits not just a more knowing and securely positioned future self who can judge but also an audience, a chain of transmission, an enduring culture that will also take up the task of deciding what to make of "these pages." It formally encodes what Povinelli describes as the liberal horizon: "an imaginary ideal and normative point toward which liberalism should be oriented and against which it should measure its state of affairs."[11] Atwood's protagonist, meanwhile, still operates according to David Copperfield's logic of deferred self-assessment, while also knowing that the time is up, the imaginary point from which things are to be measured has arrived, and that the state of affairs is disastrous and always was. The liberal horizon, the mechanism of disavowal, has arrived and it is the event horizon; on the other side lies the black hole of a fully extracted and exhausted planet. Povinelli, drawing on the work of the anthropologist Ali Feser, writes about Rochester, New York, where the operations of the now-defunct Kodak factory have thoroughly poisoned the air and groundwater with toxic chemicals, but the residents, knowing this well, and in some cases experiencing it in their bodies, nevertheless still "associat[e] the signature astringent smell of photographic chemicals with better days, happier moments, more secure futures. The sensory history of chemicals seeps into affect, creating bonds of desire, nostalgia, and mourning for the very toxins now slowly overheating bodies and landscapes."[12] Snowman, looking out to sea in the opening pages, feels the same mystifying bonds overriding the same knowledge: "On the eastern horizon there's a greyish haze, lit now with a rosy, deadly glow. Strange how that colour still seems tender."[13] In this way, *Oryx and Crake* mischievously turns the teleological drive of the bildungsroman against itself, showing how the "veil," intrinsic to the very dynamics of self-representation and a posited future integration, makes for an endless dithering that conceals the actual movement of disintegration.

In all three novels, but especially the first, Atwood uses the Crake character as diagnostic instrument to anatomize the self-fueling, self-consuming quality of denialism in which evolutionary reality and mass-marketed denial are joined in an Ouroboros-like death spiral. His is the voice of raw Darwinism: the perspective that insists on framing the human always in terms of population dynamics, aggregated consumption patterns, planetary, and evolutionary scales:

> "Jimmy, look at it realistically. You can't couple a minimum access to food with an expanding population indefinitely. *Homo sapiens* doesn't seem to be able to cut himself off at the supply end. He's one of the few species that doesn't limit reproduction in the face of dwindling resources. In other words—and up to a point, of course—the less we eat, the more we fuck."
>
> "How do you account for that?" said Jimmy.
>
> "Imagination," said Crake. "Men can imagine their own deaths, they can see them coming, and the mere thought of impending death acts like an aphrodisiac. A dog or a rabbit doesn't behave like that. . . . Human beings hope they can stick their souls into someone else, some new version of themselves, and live on forever."
>
> "As a species we're doomed by hope, then?"
>
> "You could call it hope. That, or desperation."
>
> "But we're doomed without hope, as well," said Jimmy.
>
> "Only as individuals," said Crake cheerfully.[14]

In Crake's view, biological imperatives structure all of human behavior, while the culture industry functions, as we have seen, to structure ignorance about those imperatives by offering the illusion that one day they can be controlled, outwitted, and transcended. On some level, everyone knows this dream of transcendence is a lie; still, everyone continues believing—or "believing"—in it anyway. Atwood takes a wicked pleasure in the infantile names she gives the companies pushing such fantasies—"AnooYoo," for example—as she does in detailing the symbiotic denialist logic in which culture and biology are interlocked. The "hope" that Tennyson kept behind the veil has become a cash cow at scale, in the process ironically turning the dreaded prospect of human extinction from unsettling thought to emergent reality. The biological urge to avoid death, to survive and reproduce, has metastasized into an economic and cultural imagination machine dedicated to offering endless sexual possibilities and escape from death; meanwhile, such illusions work by concealing the very biological basis of human behavior from which they sprung in the name of agency, uniqueness, and self-expression. Crake is the Grand Inquisitor of transhumanism: the unbeliever who understands better than anyone how the system of belief works and masterfully handles its controls to lead everyone to "death and destruction," as Dostoevsky puts it, via a promised bliss.[15]

The encompassing irony of Atwood's trilogy, then, is that despite all of the open discussion of evolutionary theory, despite the endless genetic experimentation and bio-hacking of the human organism, despite the fact that the entire world is very obviously facing the grim pressures of selection and survival thanks to continued environmental degradation, Darwinian evolution *still* remains an object of denial. In *The Year of the Flood*, the leader of the religious sect God's Gardeners, Adam One, decries "those who arrogantly persist in evolutionary denial," referring to the reactionary deniers, still hard at it in the distant-ish future.[16] But Atwood's main interest is the persistence of the soft denial of Darwin, which she voices through Jimmy/Snowman and his attempts to mount some kind of defense against Crake's insistence on biological reductionism. When the latter describes love as mere mating behavior dictated by the logic of sexual selection, Jimmy responds, ineffectually: "Think of all the poetry—think Petrarch, think John Donne, think the *Vita Nuova*. . . . Images, words, music. Imaginative structures. Meaning—human meaning, that is—is defined by them. You have to admit that."[17] As we have seen so often in the preceding pages, here is the predictable, almost reflexive, appeal to the achievements of human culture—specifically, familiarly, Western Culture—to attest to a sphere of meaning set apart from the merely animal. The word "admit" is Jimmy's attempt at table turning, suggesting that it is Crake who would be the denier if he refuses to acknowledge the power of such staggering testaments to human love. But clearly this is culture as fetish, poetry held up just like Conrad's "idea," as something to which everyone is expected to genuflect. Jimmy does not actually *believe* in the deeper system of spiritual significance from which such expressions of love once arose; he gestures at these works not so much because of what they mean but because they conjure a sense of *meaningfulness*. No trouble then, for Crake to breach such defenses: "People can amuse themselves any way they like. If they want to play with themselves in public, whack off over doodling, scribbling, and fiddling, it's fine with me. Anyway, it serves a biological purpose. . . . The male frog, in mating season . . . makes as much noise as it can."[18] The highest of high culture offers no portal into a transcendent realm of human experience but merely marks another two-step in the endless dance of what Crake calls "hormone robots."[19] When Crake describes his wish to alter the human mating process by making it "cyclical and also inevitable, as in the other mammals," Jimmy objects: "Think of what we'd be giving up. . . . There'd be no free choice."[20] But Crake's point is that the sense of free choice is *already* entirely epiphenomenal, something that only disguises the deeper, impelling biological dictates of survival and reproduction. Jimmy *knows* this is the case but *feels* it to be otherwise, and in this he is meant to represent, in the early pages at least, something of the Western bourgeois everyman: just insulated enough to be susceptible to the fantasy in spite of himself.

No one wants to be a "hormone robot." No one wants to think of the human story as "an experiment animal protein has been doing on itself" either.[21] The novels are filled with such descriptors that reduce the human to its biological functions, along with casually brutal scenes in which the human form literally loses its shape and integrity. A security guard is "dissolved into a puddle of goo";[22] someone is thrown over an overpass and becomes "car tire mush";[23] another catches a bioengineered hemorrhagic fever that turns him into "a raspberry soda";[24] and a fast-food franchise is suspected of turning human carcasses into hamburger meat.[25] As we have seen in *The Time Machine, Heart of Darkness, Dr. Jekyll and Mr. Hyde,* and even in unexpected places like *Daniel Deronda,* the nineteenth-century novel was a place where the image of the human was simultaneously stabilized and destabilized; indeed, at times, it was a place where the clash of those opposing tendencies could be dramatized, where the wish or need to maintain a coherent picture of the human could be set against the forces that were everywhere working to undo it. To return once more to Ian Duncan, nineteenth-century literature served as a zone of experimentation, working hand in hand with the natural sciences to simultaneously break and fortify the boundaries of the human:

> Novels could offer a comprehensive representation of human life—a Human Comedy—in a general writing accessible to all readers, mediated not by spe-

cialist knowledge or technical language but by the shared sensibilities that constitute "our common nature." Novels became active instruments in the ongoing scientific revolution, advancing its experimental postulates—that human nature may not be one but many, that humans share their nature with other creatures, that humans have no nature, that the human form is variable, fluid, fleeting—as well as developing a technical practice, realism, to defend humanity's place at the center of nature and at the end of history.[26]

Advancing *and* defending: the antinomies pushed to extremities in Atwood's trilogy mark the terminus of this Human Comedy and of the comic mode itself, which functions as instrument of critique, source of pleasure, and agent of undoing all at once. I noted in the Introduction that disavowal often has a kind of familiar comic structure: *I know this is bad for the environment, but I like it. I know I shouldn't ignore the climate crisis, but I do anyway.* Such social exchanges mark a nervous reaffirmation of a shared sensibility, the ratification of "our common nature" and the boundaries of the human reassuringly (if ironically) scaled to the everyday. If we can still joke about it, it must not be that bad. In Atwood's hands the novel continues to serve this function, while it also makes plain the cost of framing the human story *as* a comedy, of treating serious things unseriously in a defensive reconsolidation of the human in response to gathering threats to its existence.

All of these questions meet in the protagonist's post-apocalyptic *nom de guerre*, Snowman, which is how he introduces himself to the "Crakers," the lab-born, genetically engineered members of Humanity 2.0: "It's given Snowman a bitter pleasure to adopt this dubious label. The Abominable Snowman—existing and not existing, flickering at the edges of blizzards, apelike man or manlike ape. . . . For present purposes he's shortened the name. He's only Snowman. He's kept the *abominable* to himself, his own secret hair shirt."[27] Jimmy the chronic jokester has become the joke: an in-joke, only to himself. Whatever pleasure there was in humor has turned bitter, as it reinforces his sense of isolation and entrapment in the human nature he believes he now shares with no one else alive. By novel's end, Snowman is limping along with a cane because of a badly infected foot, one of many ironic callbacks to Sophocles' *Oedipus Rex* and a nod to the centrality of the human form in that story. *Oedipus* is, at heart, about the painful process of enlightenment emerging from a state of denialism; here, Atwood suggests a rewriting of the script as well as the absurdist conclusion to the humanist project it inaugurates. To the Sphinx's famous sequence of silhouettes marking the human life cycle—four legs, two legs, three legs—Atwood offers one more: a "Snowman." No legs. After the hobbling comes the melting. Lol. This is an in-joke, a secret bit of comedy for the reader in the know, laced with unmistakable bitterness. The age of human primacy is followed by obsolescence, a world of disappearance, extinction, breakdown. On an overheated planet, the word "snowman" has come untethered from the environment in which it made sense, in which the human could be given shape, even a pretend one.

The Darwinian world manufactures the denial of the Darwinian world. This is what Atwood anatomizes, the way the conceptual and material disintegration of the human form, always and everywhere on display, makes those commodities, practices, and agencies that pretend to put it back together feel only more, rather than less, necessary. One of the most terrifying aspects of *Oryx and Crake* is the sense of enclosure it creates through this logic, the way everything becomes gradually sealed, including Atwood's text itself—its jokes, its narrative procedures, its imaginative daring—into the hermetic logic of a totalized field. This includes the third-person narrator whose masterful use of the ironies of free indirect discourse provides a knowing, entertaining indictment of a culture of entertainment and knowingness. *Oryx and Crake* itself is thus positioned as both diagnosis and symptom, agent of critique and complicit participant in the dynamics of the system it represents. Something of this is registered in the choice of the first epigraph for the novel, from another work about human form-bending, Jonathan Swift's *Gulliver's Travels*: "I could perhaps like others have astonished you with strange improbable tales; but I rather chose to relate plain matter of fact in the simplest manner and style; because my principal design was to inform you, and not to amuse you."[28] Amusingly disavowing the intention to amuse by referencing a text that itself amusingly disavows the intention to amuse—a hall of mirrors is diverting, but it also hides the exit. And reflected in it is something of Atwood's more serious intention in referencing Swift, which is channeling the undercurrent of rage, despair, and comprehensive disgust at the human world that flows through his work straight into *Oryx and Crake*.

The novel's second epigraph is from *To the Lighthouse*: "Was there no safety? No learning by heart the ways of the world? No guide, no shelter, but all was miracle and leaping from the pinnacle of a tower into the air?"[29] As we've seen, Woolf's novel, like Atwood's, is also about the relationship between the forms of ignorance through which the human world is structured and the inevitable collapse of the material structures of a world so constituted. This specific passage, so divergent in tone and import from Swift's, pairs with it to suggest the divided imaginative landscape of *Oryx and Crake* itself. On the one hand, the comprehensive, closed-system satire of the Enlightenment project and the hopelessly deluded human animal at the center; on the other, the more open possibilities of the novel form at its most boundary pushing, where something new and miraculous still might emerge. Of course, as Atwood is well aware, the valorization of the modernist novel *as* a boundary-pushing device sits uneasily in this world of experimentation run amok, and the leap of faith off the tower might have a quick end when it meets the law of gravity. The point of putting Swift and Woolf together like this is not to counterpose dystopia and utopia but to put the dystopic in touch with what Fredric Jameson calls "utopian praxis"—the urgent demand to imagine the new rather than the exact specification of what that new should consist of: "It is less revealing to consider Utopian discourse as a mode of narrative, comparable, say, with novel or epic, than it is to grasp it as an object of meditation, analogous to the riddles or *koan* of the various mystical traditions, or the aporias of classical philosophy, whose function is to provoke a fruitful bewilderment, and to jar the mind into some heightened but unconceptualizable consciousness of its own powers, functions, aims and structural limits."[30] That kind of mind-jarring bewilderment is where Woolf lives and why Atwood invokes her at the outset. We're all, collectively, leaping off the tower anyway, whether we like it or not. Perhaps in the recognition of that resides the conditions for a different kind of social imaginary to come into being. It's not like there are great alternatives.

The *MaddAddam* trilogy returns us in important ways to the issues we began with in the opening sections of this book: Dillard's turning away in loathing from the insect world; Charlotte Brontë's refusal of Martineau's scientific materialism; George Eliot's questions about what should and shouldn't "enter" human consciousness. As I have said before, the climate crisis indicates that turning away in horror from material reality, treating it as if it could be kept "purely external," is a recipe for disaster. Atwood seems to think that too. But she also seems to think that questions remain about what the desired knowledge position actually *is*. That is, it is one thing to expose denial as a massive cultural problem; it's another thing to define its opposite. Not to be in denial is to be what, exactly? Fully aware, all the time? Totally disillusioned? Is such a condition desirable, if it were even possible? And isn't the idea of achieving such a position also a fantasy of the sovereign, perfectly knowing subject? We are back on the horns of Freud's dilemma. In Atwood's trilogy, especially in the first book, Crake would seem to represent this fantastical state: He's a Dostoevskyean thought experiment, a deliberately unreal character whose life is dedicated to imagining the human as a biological organism, stripped of all pretensions, illusions, and consolations. This includes, it would seem, himself. We never really know what Crake is thinking, but he arranges for his own elimination at Jimmy's hands, acknowledging, apparently, that his own humanness means that he too must be removed from the scene in order for the new world to come into being. Jimmy's mother admires Crake for his willingness to follow thought wherever it leads and his utter lack of self-deception: "You could have an objective conversation with him, a conversation in which events and hypotheses were followed through to their logical conclusions. . . . 'Your friend is intellectually honorable,' Jimmy's mother would say. 'He doesn't lie to himself.'"[31] Jimmy, meanwhile, is selected to survive by Crake for the very half-in, half-out quality that will make him a cynically detached but also grudgingly committed caretaker of the new species. *Fluent in disavowal* could almost be a line on his resume, the practiced skills of the company man that make him the ideal middle manager for Crake's new enterprise.

Of course, Atwood would have us register the dark irony here that this "intellectually honorable" state of total non-denialism, Crake's unblinkered view of humanity on a species level and planetary scale, leads him to become the most prodigious mass murder in all of human (and natural) history. Anthropocidal rather than merely genocidal. Here Atwood also shows herself working along the fissure opened in the nineteenth-century imaginary, which Charles Taylor describes as "split-screen vision of nature" brought about by the development of geology and then biology: "This is one respect in which our cultural predicament is utterly different from what existed before the eighteenth century, where the scientific explanation of the natural order was closely aligned with its moral meaning. . . . For us, the two have drifted apart, and it is not clear how we can hope to relate them."[32] This sense of drifting apart is Atwood's subject in all three novels, conspicuously allegorized in the marital split between Jimmy's parents in the first book, which, we are to understand, produces the divide within himself and thus his state of terminal indecision, paralysis, and denialism. Their representative function is made clearest in his mother's horrified reaction to his father's research into growing human neo-cortex tissue in pig brains for the bioengineering firm NooSkins: "'There's research and there's research. What you're doing—this pig brain thing. You're interfering with the building blocks of life. It's immoral. It's . . . sacrilegious."[33] To which his father reacts: "'Who've you been listening to? You're an educated person, you did this stuff yourself! It's just proteins, you know that! There's nothing sacred about cells and tissue, it's just . . .'"[34] Here is the predicament in a nutshell: there is a felt sense of a profound trespass but no available vocabulary or coherent moral structure upon which to mount an objection. The idea that there is "nothing sacred" is the deep problem these novels worry over. There are no boundaries capitalism will not steamroll in the hunt for profit; but there is also no way to resacralize the natural world, or the human, to even attempt to establish some line of defense. Jimmy's mother reaches for language that no longer has any grip on reality, just as Jimmy, as Snowman, will find himself obsessing over a suddenly replete roster of obsolete words after the apocalypse has obliterated the social context to which they had been attached. "Toast" brings this dynamic to a bleakly ridiculous climax in the magnificent mini-chapter of that name. Like Jimmy's references to Petrarch, Donne, and Dante, his mother uses "sacrilegious" and even "immoral" fetishistically, as signifiers no longer attached to the world of transcendent signifieds from which they once derived their power, but she wields them *as if* they still have it. To see the human mind as "just proteins" is to think like an ecologist. But it is also to think like a capitalist. And here is the rub: the language that might be used to push back against environmental spoliation and human degradation has either been already co-opted by capitalism or just doesn't mean anything anymore.

Atwood's novels thus offer a scathing and thoroughly imagined critique of a culture of denial produced by capitalism, but they also address themselves to the other side of the issue, which is the sense that some kind or some amount of denial seems necessary for both individual well-being and a nonexploitative form of social functioning. Not to be *well* wadded with stupidity, but *kinda* wadded. The question Atwood struggles with in all three books (and maybe the reason why there *are* three books) is the one George Eliot also struggles with—lines must be drawn, but where, exactly? And according to what principle? And by whom? How much wadding is too much? And what kind of stupidity is the right kind? Recall John Kucich's argument that Eliot was "appalled by man's universal and irreversible amalgamation with matter."[35] The point is that, for all the alacrity with which she sets about puncturing any and all forms of human vanity, Atwood is too. Or, to put it more precisely, she is appalled by the uses to which such amalgamation can get put: the reduction of all living and nonliving things to a potential source of extractible profit as capital works to vacuum every last bit of value out of the planet and its inhabitants. As Achille Mbembe puts it, "The degree to which capital today is adept at exploiting this constitutional consumability of beings and species represents a major inflection point in the history of humanity. It radically redefines the very nature of 'the human' and forces us to revisit the categories by which we used to conceive of social life."[36] Or, Atwood suggests, forces us to avoid doing so until it is too late.

The second two books of the trilogy are devoted to this question, and the shift in attitude toward denialism is signaled by a change in protagonist. *Oryx and Crake*, organized around Jimmy/Snowman, offers a demolishing critique of his jokes and defense mechanisms and structures of ignorance. *The Year of the Flood* and *MaddAddam*, meanwhile, both find their center of gravity in the more sympathetic figure of Toby, another survivor of the apocalypse and another inveterate soft denier. Indeed, there is something continuous about the voicing of all three novels, insofar as Atwood goes out of her way to show us, through the operations of free indirect discourse, how Toby, like Jimmy, dwells in divided states of knowing and not knowing: "It doesn't bear thinking about, so she won't think about it. *Will not* think. Wills herself *not*";[37] "Surely she'd buried that particular sadness many years ago. If not, she ought to bury it";[38] "She blots out the pictures. Or she tries to blot them out. . . . A vortex opens: she closes it";[39] "She tries not to think of the other things [the vultures] dropped in the few weeks after the Flood. Fingers were the worst";[40] "Toby suppressed the memory of eating these burgers."[41] And yet, where *Oryx and Crake* takes a dark delight in showing the protagonist's denialism getting shredded by reality, the second two novels treat Toby's denial as a resource, a coping strategy, and then ultimately as a way of rebooting belief amid the ruins of capitalism. In *The Year of the Flood*, she is rescued by a group of religious believers called God's Gardeners, whose mission is to address the predicament Taylor discusses and reinvest the natural with a sense of moral and spiritual meaning. Toby's relationship to the movement is characterized by a mixed condition of knowing and not knowing at once. "Nature never does betray us. You do know that?" one of the Gardeners asks, to which the narrator tells us: "Toby knew no such thing."[42] And yet, she follows their practices, murmurs their prayers, and believes enough in their mythology to talk to bees as if they can carry messages to the dead: "Had she believed all that? Old Pilar's folklore? No, not really; or not exactly. Most likely Pilar hadn't quite believed it either, but it was a reassuring story: that the dead were not entirely dead but were alive in a different way; a paler way admittedly, and somewhere darker. But still able to send messages, if only such messages could be recognized and deciphered. People need such stories, Pilar said once, because however dark, a darkness with voices in it is better than a silent void."[43] "Better" for the individual person and better for the social world. The spiritual leader of the Gardeners, Adam One, sounding a bit like Dmitri Karamazov, stresses the need for belief, whether you believe it or not: "We must be a beacon of hope, because if you tell people there's nothing they can do, they will do worse than nothing."[44]

The remaking of the world involves refashioning disavowal into a creative rather than a defensive act. Life itself becomes something in which one must suspend one's disbelief: "She didn't really believe in their creed, but she no longer disbelieved."[45] Such a position skates perilously close to the divided mind that helped produce this mess in the first place, just as the return of organized religion, however gentle its founding creed, must give us pause. We've seen this story before. The idea that new modes of belief and social organization might rise out of the wreckage of capitalism is personified by Adam One, a character who emerges from an abusive household run by his televangelist father, the leader of the (somewhat on-the-nose) "Church of PetrOleum."[46] Adam One clearly serves as Crake's benevolent double in the narrative economy: an aloof, designing character whose inner life remains opaque, whose true feelings are only to be inferred through his deeds and sayings, and who surveys the human situation from a superior remove. The man behind the curtain—or, perhaps we should say, the veil—never fully known, working upon the world in mysterious ways. *Deus absconditus*: the god who makes and splits. Through the dual workings of these secular demiurges, Crake and Adam One, the great hope for the future comes to reside in the genetic merger of the remaining humans and the bioengineered Crakers, the hybridized offspring of which are only glimpsed in the conclusion but whose evolutionary path seems to promise a better way forward. It's a happy-ish ending, made possible by what is essentially an extended deus-ex-machina plot, the consoling design of which really only reveals itself at the end. It is "apocalyptic" in the original, Greek sense, which means there is also a switch, at trilogy's end, from an interest in the reconstructive powers of fictionality within the diegetic world of the novel to the reconstructive powers of fictionality expressed through the design of the plot itself. It's an expression, I would venture to say, of Atwood's own wishfulness, her own desire to believe or suspend disbelief made possible through, given form by, the artifice of fiction. This is something Atwood herself no doubt understands well. It is, in a sense, a reboot of providential plotting, which we are invited to both to take seriously and not, to feel powerfully and recognize as a trick, a formal convention. And thus do we find ourselves, inevitably, back with the Victorians.

"Let us hasten to cast a veil over these horrors," writes the nineteenth-century French entomologist Jean-Henri Fabre after describing an especially gnarly scene from the life of insects.[47] Darwin references Fabre in both *The Origin of Species* and *The Descent of Man* and corresponded with him in early 1880, complimenting his powers as a "wonderful observer" (always high praise from CD) and commending his book *Souvenirs entomologiques*, parts of which were later translated as *The Hunting Wasps*. He would make good use of this book, Darwin tells Fabre, "if I were to write on the evolution of instincts."[48] Fabre's book was also at hand at Tinker Creek, and Dillard uses it to process her own response to the insect life she observes with fascination and loathing. "The remarkable thing about the world of insects," she writes, "is precisely that there is no veil cast over these horrors. These are mysteries performed in broad daylight before our very eyes; we can see every detail, and yet they are still mysteries."[49] But when "horrors" is changed to "mysteries," is the veil not returning at the very moment we're told it isn't there? Is this the denial of denial and therefore just more denial? *Pilgrim at Tinker Creek* wonders about this too and then wonders, further, whether its own recursive structure and intricate layers of reflection and revisiting are a way to know or just another complex means of not knowing.[50] Like her predecessors Emerson and Thoreau, Dillard mistrusts self-consciousness, foregrounds it as a problem, and yet frequently finds herself formally encoding its processes as she attempts to escape from them. Like Atwood, she confronts the world of Darwinian horrors, feels the force of a need for something beyond the basely material, and distrusts it for that very reason. While Dillard's organizing concerns are spiritual-existential, and Atwood's socio-political, both writers build out of their reaction to Darwin complex, self-assessing formal structures that artfully expose, as they artfully express, the workings of denial.

Fabre, it should be said, was an evolution denier. Today you can find him profiled admiringly by a depressingly familiar mix of zealots and con artists on various creationist websites. With characteristic tact, Darwin registers his disappointment with Fabre's position: "I am sorry that you are so strongly opposed to the Descent theory," he writes, before gently suggesting that an evolutionary perspective would open up new possibilities for his research: "I have found the searching for the history of each structure or instinct an excellent aid to observation; and wonderful observer as you are, it would suggest new points to you."[51] For my part, I find Fabre a sympathetic figure for the way he dedicated his entire professional life to studying something he didn't want to look at too closely. Fabre ends *Souvenirs entomologiques* with a dedication to his son, Jules, who died at age sixteen before the book was finished: "Hélas! tu es parti pour une meilleure demeure, ne connaissant encore du livre que les premieres lignes! Que ton nom du moins y figure, porté par quelques-uns de ces industrieux et beaux Hyménopteres que tu aimais tant."[52] In the folklore Fabre knew so well, bees and wasps often play the role of messengers between the lands of the living and the dead. It is the same mythology Atwood's surviving characters agree to try to semi-believe in. Like Fabre, confronting unimaginable loss, they imagine "a darkness with voices in it," hoping to feel whatever consolation such sounds might offer. Having suffered the loss of children himself, including his beloved daughter, Annie, only ten, Darwin closes his letter to Fabre in the present tense of grief, one bereft father to another: "Permit me to add that when I read the last sentence in your book, I sympathise deeply with you—"[53]

Darwin never wrote his book on the evolution of instincts (he died two years after writing this letter), but Hymenoptera, which Fabre observes so wonderfully, plays a complicated role in both *The Origin of Species* and *The Descent of Man*. Fabre's wasps are closely related to the parasitic Ichneumonidae whose horrifying behavior so famously and fatally disturbed Darwin's belief in a benevolent Creator, as he writes to Asa Gray: "There seems to me too much misery in the world. I cannot persuade myself that a beneficent & omnipotent God would have designedly created the Ichneumonidæ with the express intention of their feeding within the living bodies of caterpillars."[54] At moments in *The Origin of Species*, bees and wasps arise like creatures from a nightmare, implanting larvae inside living animals, murdering their own offspring, challenging the reader's ability to stomach the theory while providing vivid illustrations of how it works.[55] At other moments, they allow Darwin to strike at the design argument where it lives. If natural selection can explain how bees can make hives of such intricacy and geometric perfection, then a major plank in the creationist platform goes to splinters.[56] At still others, they play starring roles in his ecological storytelling—not so much as messengers between living and dead but as "indispensable" connectors invisibly uniting a range of organisms throughout the web of life.[57] In chapter 3, he describes how bees pollinate red clover, which means that humans, who own house cats, which keep mice away, which otherwise attack bee hives, help keep the clover in bloom.[58] There is grandeur in the evolutionary view, Darwin famously says at the end of *The Origin*, but this scene, more humble than grand, is no less compelling for that. It paints a modest picture of mutual relationality among what Donna Haraway calls "companion species": "such organic beings as rice, bees, tulips, and intestinal flora, all of whom make life for humans what it is—and vice versa."[59] To this idea of companionability, we might add what Carla Hustak and Natasha Myers call "affective ecologies," which include "creative, improvisational, and fleeting practices" as well as new forms of communication among "articulate plants and other loquacious organisms."[60] Such new ways of imagining nonverbal interspecies communication and non-hierarchical relationality are what drive Atwood's survivors and the radically new human world they are trying to structure for themselves, even if they only half believe in it.

We know today that humans have been less-than-companionable companions to the bees. Bee populations are crashing worldwide thanks to (among other things) the indiscriminate use of pesticides, the effects of which Rachel Carson warned us about decades ago: "The chemical destruction of hedgerows and weeds are eliminating the last sanctuaries of these pollinating insects and breaking the threads that bind life to life. These insects, so essential to our agriculture and indeed to our landscape as we know it, deserve something better from us than the senseless destruction of their habitat."[61] We also know that ichneumon wasps are "especially sensitive to pesticides" and are powerful predators that target leafroller moths, wooly apple aphids, boll weevils, and other invasive species, which means the senseless destruction of these wasps, who certainly do deserve better, also might be the senseless destruction of a better way of protecting an agriculture and landscape that have also been put at risk.[62] Perhaps once again thinking of Darwin, his lost faith, and his fears, Dillard writes about the ichneumon wasp in *Pilgrim at Tinker Creek*. Instead of focusing on the gruesome results of its parasitical behavior, she turns her attention to the impossible dilemmas facing a mother wasp when she can't locate a host for her young. Should she jettison them before they hatch, leaving them to die, or should she let them hatch and devour her? If the wasp was making conscious choices, Dillard, writes, "Her dilemma would be truly the stuff of tragedy; Aeschylus need have looked no further than the ichneumon."[63] But, of course, the idea is that the wasp is *not* making conscious choices, so not to worry—if it's tragic material we're after, Aeschylus and the rest of us need only keep our attention fixed upon the human world. But then, as often happens in this book, the terms suddenly change and the humanist ground shifts underfoot: "That is, it would be the stuff of real tragedy if only Aeschylus and I could convince you that the ichneumon is really and truly as alive as we are, and that what happens to it matters. Will you take it on faith?"[64] Hailed unexpectedly, the reader is left to wonder what she is being asked to agree to. Who would dispute that a wasp is as "alive" as we are? And yet, in another sense, who would actually agree to it? What happens to another species obviously *matters*—we all know we're supposed to say that. And yet, just as obviously, we have constructed an entire system of productive practices, daily behaviors, habits of attention, strategies of evasion, aesthetic conventions, which, taken as a whole, suggests that we, in fact, *don't* really think they matter. To imagine the freedom of the wasp, then, is to imagine the possibility of freedom from our own patterns of mindless reproduction, from the demand to tend to the needs of a system bent upon our own destruction and built upon the denial of life itself. The grass grows, the squirrel's heart beats, the wasp wonders what to do next—Dillard's voice projects into the other side of silence with archness and sincerity in equal measure. It strikes the deeper chords of a common animal being as it plays along the surface of make-believe, remixing modes of knowing and feeling, thinking and imagining, in open and unresolved tones, all the way to the end. Today, a half century after she wrote this, a full fifty years deeper into a planetary tragedy of our own making, we cannot help hearing in her prose the unmistakable notes of warning about the wages of a stunted and impoverished ecological imagination and the tragic outcome of a long refusal to admit that everything really and truly is as alive as we are, and what happens to it matters.

NOTES

Introduction

1. Dillard, *Pilgrim at Tinker Creek*, 176.
2. Darwin, *The Origin of Species*, 65.
3. Although psychoanalysis is still some decades away, these terms—"deny" and "admit"—have already begun to accrue something of their more psychologically fraught meaning. As Catherine Hall and Daniel Pick argue, this shift can be traced back as far as the eighteenth century: "Denial, even in the specific sense of the quelling of an internal conflict, is not some exclusively modern concept, nor the preserve, alone, of Freud and his followers. Johnson's *Dictionary* of 1755 glossed denial to mean negation, refusal, or even abjuration; the latter defined as the contrary of an acknowledgment of adherence. Johnson also included an entry for the term 'denier,' meaning a contradictor, an opponent, one that holds to the negation of a proposition, but also potentially a 'disowner,' 'one that does not own' or acknowledge, or even a 'refuser,' 'one that refuses.'" "Thinking about Denial," 9.
4. Norgaard, *Living in Denial*, 5.
5. Cohen, *States of Denial*, 8.
6. Mannoni, "I Know Very Well," 68.
7. As Klein puts it, "We focus too much on climate deniers and not enough on the more widespread 'soft denial.' How is it possible to know about this crisis, then forget? What is all this aversion about—how can we know something so profoundly disturbing and then behave as if it isn't happening?" MacDonald, "Naomi Klein on Climate Change."
8. Latour, *Facing Gaia*, 9.
9. Dillard, 64.
10. Sedgwick, *The Weather in Proust*, 212, 208.
11. Sedgwick, 210.

12. Sedgwick, 212.
13. Sedgwick, 209.
14. Dickinson, "The People Paradox."
15. Becker, *Escape from Evil*, 85.
16. Tuan, *Escapism*, xiii.
17. Fowles, *The French Lieutenant's Woman*, 97.
18. Tuan, xvi.
19. Latour, "Will Non-humans Be Saved?," 468.
20. Gould, *Full House*, 19.
21. Gould, 29.
22. Gould, *Ever Since Darwin*, 37–38.

23. As the semiotician Paul Bouissac argues, this would include the human relationship to non-human animals. In his critical review of the edited collection *What Is an Animal?* he takes issue with the category "animal" as it is commonly used in philosophy and anthropology, arguing along lines very similar to Gould's: "Even if their authors pay lip service to evolutionary theory (e.g., Clark) or state their disagreement with Descartes' extreme position (e.g., Midgley), they all appear to be concerned primarily with the defense and illustration of man's uniqueness in the universe. This assumption is so much an integral part of the current Western world-view and system of values that the mere uttering of the question 'What is an animal?' can only emanate from this world-view (which holds that there is such a cognitive category) and at the same time cause a deep anxiety, since the question implies that one might overlook or even ignore the dividing line which negatively defines humans. Darwin's theory and its subsequent developments are still scandalous for a culture still firmly established upon markedly different principles." "What Is a Human?," 505.

24. Gould, *Full House*, 19.
25. Wilson, *On Human Nature*, 209.
26. C. Taylor, *The Sources of Self*, 406.
27. Wilson, 201.

28. Seen in this light, Wilson is not exactly being inconsistent in his picture of human nature, since his view of biological reductionism and his argument for attending to the affective "needs" of people are, on some level, describing different aspects or spheres of the human experience, one of which is founded on necessary illusion. There is an implicit divide between a strictly scientific view of *Homo sapiens* and a social or political imaginary that, he thinks, depends upon "mythology" and the production of "blind hope" and puts the idea of the human to purposes other than accurately describing the organism in material terms. And yet, we can also see how entirely unstable this situation is, since it is itself a fantasy to believe the scientific can be cordoned off from the social or the political.

29. C. Taylor, 416–17.
30. Wilson, 207.
31. See MacDuffie, *Victorian Literature*, especially 43–49.

32. Leaving aside the (newly reopened) question of whether Wilson's evolutionary ideas were structured by white supremacist ideology, it is beyond dispute that they have

been eagerly taken up by various racist and eugenicist causes, and there are at least some clear genealogical links between modern sociobiology and what once was known as Social Darwinism. For a discussion of these links, see Mary Midgley's "Selfish Genes and Social Darwinism" and Janna Thompson's "The New Social Darwinism."

33. Wynter, "Unsettling the Coloniality of Being," 260.
34. Chakrabarty, "The Climate of History," 219.
35. Karera, "Blackness and the Pitfalls of Anthropocene Ethics," 32.
36. Karera, 50.
37. Wilson, *The Future of Life*, 79.
38. Chakrabarty, 215.
39. Wilson, "Foreword" to Sachs, *Common Wealth*, xii. Quoted in Chakrabarty, 215.
40. Chakrabarty calls Wilson's solution "quasi-Hegelian" in its emphasis on the process of coming to self-knowledge (215). As we'll see, the valorization of self-consciousness was a key hinge in the disavowal of the ecological implications of Darwin's work.
41. Carson, *Lost Woods*, 245.
42. Carson, 210.
43. Indeed, George Levine's *Darwin and the Novelists* and Gillian Beer's *Darwin's Plots*, both published in the 1980s, are as canonical as canonical gets in the field of Victorian studies, and books in the last twenty or so years by Cannon Schmitt, Gowan Dawson, and Devin Griffiths have all brilliantly followed Beer and Levine in exploring the manifold connections between evolutionary biology and literature in the long nineteenth century and beyond.
44. Morton, *The Ecological Thought*, 18.
45. Grosz, *Becoming Undone*, 13.
46. Griffiths, "The Ecology of Form," 72–73.
47. The "animal turn" often features Derrida and his late work *The Animal That Therefore I Am* rather than Darwin, although Colin Nazhone Millburn and Philip Barrish both make the case for Darwin as an unacknowledged forerunner of deconstructive thought. As Barrish puts it, "Darwin's account of the origin of organic species prefigures a wide range of theoretical models that have recently become influential in the reading of literature." "Accumulating Variation," 431.
48. Duncan, *Human Forms*, 14–15.

Chapter One

1. Malm, *White Skin, Black Fuel*, 17.
2. Marx had complicated thoughts about Darwin: he admired his work for its materialism and its rejection of teleological explanations of nature but also worried that it naturalized capitalism. He wrote to Engels: "It is remarkable how among beasts and plants Darwin rediscovers his English society with its division of labor, competition, opening up of new markets, 'discoveries' and Malthusian 'struggle for existence.'" "To Engels," 380.
3. Gosse, *Omphalos*, 103–4.
4. "Review of *Omphalos*," 55.
5. Hansson, "Science Denial," 42–43.

6. Wilberforce, "Darwin's *Origin of Species*," 239.

7. For Darwin's comment on Wilberforce, see Desmond, *Huxley*, 283. The phrase "merchants of doubt" is from Oreskes and Conway, *Merchants of Doubt*.

8. "Jefferey's British Conchology," 90.

9. "Geology and Protestantism," 379. For this and the following two references, I am indebted to Alvin Ellegård's marvelous *Darwin and the General Reader*.

10. "Geology and Protestantism," 380–81.

11. "Darwin and *The Descent of Man*," 195–96.

12. This is to say nothing of the openly emotive, name-calling reactions that characterized the anti-Darwinist response, like this one from the ornithologist Francis Morris: "'Ineffable contempt and indignation' is the only feeling which any person of common sense and of a right mind must feel at the astounding puerilities of Darwinism, its ten thousand times worse than childish absurdities; contempt for them in themselves and indignation at the criminal injury the miserable Infidelity of the wretched system has done to the minds of too many." *All the Articles*, 96.

13. Morton, *Hyperobjects*, 48.

14. Darwin, *The Origin of Species*, 420.

15. Levine, "The Sounds of Silence," 142.

16. Darwin, 83.

17. Tennyson, *The Major Works*, 236.

18. Morton, 47–48.

19. Levine, 142.

20. Eliot, *The Lifted Veil*, 83–84.

21. Eliot, 40.

22. Eliot, *Miscellaneous Essays*, 247.

23. Dostoevsky, *The Brothers Karamazov*, 589.

24. T. Huxley, "Geological Reform," 234.

25. Tennyson, 237.

26. Moretti, *The Bourgeois*, 109.

27. Beer, *Darwin's Plots*, 146–47.

28. For this reference, I am indebted to Moretti's *The Bourgeois*, which first made me aware of Brontë's use of this trope.

29. Brontë, "Letter to James Taylor," 574.

30. Clifford, *Seeing and Thinking*, 156.

31. Tennyson, "The Making of Man," 86.

32. Merchant, *The Death of Nature*, 143–44.

33. Wynter, 262.

34. Bateson, *Steps to an Ecology of Mind*, 491–92.

35. Wynter, 317–18.

36. Wynter, 318.

37. Fanon, *The Wretched of the Earth*, 311.

38. Povinelli, *Between Gaia and Ground*, 43.

39. Povinelli, 41–42.

40. Gould, *Full House*, 19.

41. Bilgrami, *Secularism, Identity and Enchantment*, 149.

42. The liberal re-enchantment that I'm interested in here should be distinguished from George Levine's argument in *Darwin Loves You* that despite the "conventional understanding of him as a primary disenchanter of the world ... Darwin's work can be read as contributing to a radical *re*-enchantment of the world" (22). For Levine, that contribution involves not an embrace of the liberal horizon but something akin to what Dillard is after in *Pilgrim at Tinker Creek* and E. O. Wilson terms "biophilia," an apprehension that, after Darwin, "we find ourselves in a world of wonders, a world worth loving; we become participants and observers in a life larger than any of us, and more meaningful" (25).

43. Herbert, *Victorian Relativity*, 157.
44. Darwin, *The Descent of Man*, 689.
45. Darwin, *The Origin of Species*, 74.
46. Steiner, *Psychic Retreats*, 92–93.
47. Mannoni, 68.
48. Nabokov, *Lectures on Literature*, 6.
49. Woolf, *Collected Essays*, 57.
50. Woolf, 58.
51. Woolf, 58.
52. T. Clark, 159.
53. Ghosh, *The Great Derangement*, 78.
54. Ghosh, 25–27.
55. Terry, *Anthony Trollope*, 94.
56. Ghosh, 35.
57. Darwin, *Journal of Researches*, 219.

58. Although it should also be noted, as Jesse Oak Taylor (following the historian David Sepkoski) argues, that "Darwin's commitment to gradualism and his overriding opposition to human exceptionalism led him (and even more so, others who followed in his footsteps) to misconstrue the unique intervention marked by the large-scale transformation of the biosphere that was unfolding before their eyes." "Darwin after Nature," 25.

59. Browne, *Charles Darwin*, 69–70.
60. Ghosh, 23.
61. Trollope, *Thackeray*, 185.
62. Romanow, "Metafiction as Reality Effect," 1100.

63. The phenomenon Romanow discusses is, on some level, a nineteenth-century precursor to the "we live in a simulation" or "the scriptwriters are getting lazy" joke: the sense that life is, in some sense, scripted and/or unreal and that we are not only ourselves always "performing" our identities in the Erving Goffman sense, but we also inhabit a fictional world that is being written or observed by an outside entity. We might connect this to the old question of the "Disappearance of God," discussed famously by J. Hillis Miller: the sense of real life as a "fiction" seems both a holdover from, and a reaction to, the loss of belief in an all-watching and (in some traditions) all-determining author. Here, indeed, is another sense of the "veil": the notion that the material world, including the human body, is a veil obscuring the "real" reality, the world of the spirit. According to 2 Corinthians

3:16, "Nevertheless when it shall turn to the Lord, the veil shall be taken away." Darwin fatally undermined the notion of providential narrative patterning in nature, the implications of which, Latour says, we have yet to fully absorb: "For [Darwin], there is no overall narrative, no controlling divinity. Each individual organism is alone with its own risk, goes nowhere, comes from nowhere: it is creativity all the way down" ("Will Non-humans Be Saved?," 470). Maybe so, but surely this is not how most people construct a sense of their own lives, in the nineteenth century or now.

64. Duncan, *Human Forms*, 8.
65. Duncan, 6.
66. Jameson, *The Antinomies of Realism*, 6.
67. Jameson, 11.
68. Klein, *This Changes Everything*, 3.
69. Berlant, *Cruel Optimism*, 1.
70. Peci, "Letter."
71. Marx, *Capital*, 51.
72. Marx, 580.
73. Soper, "Greening Prometheus," 87.
74. Plumwood, *Feminism and the Mastery of Nature*, 21.
75. Mulvey, "Some Thoughts on Theories of Fetishism," 12.
76. Soper, 87.
77. Seabrook, *The Race for Riches*, 101, 100.
78. Seabrook, 101.
79. Robbins, *The Beneficiary*, 5–6.
80. Žižek, "Introduction," 8.
81. Žižek, *For They Know Not What They Do*, lxxi.
82. Sloterdijk, *Critique of Cynical Reason*, 5.
83. Sloterdijk, 5.
84. Glasson, "Aware but Unmoved," 15.
85. Samalin, "Introduction," 426–27.
86. T. Clark, *Ecocriticism*, 159.
87. Poovey, *Making a Social Body*, 25.
88. Poovey, 26.
89. Benjamin, *The Arcades Project*, 389.
90. Benjamin, *One-Way Street*, 89.
91. Arendt, "What Is Freedom?," 145.
92. Arendt, 146.
93. Arendt, 145.
94. Didion, *The White Album*, 11.

Chapter Two

1. Darwin, "To Emma Darwin," 84.
2. Phillips, *Darwin's Worms*, 37.
3. Darwin, "To J. D. Hooker," April 7, 1859, 280. His letters, especially to Hooker, repeatedly testify to this link between the feelings of the body and a kind of bleak cosmic

Notes to Chapter 2

picture. See, for example, his letter of May 31, 1866: "I should much like to be amused, for my stomach & the whole Universe is this day demoniacal in my eyes" (194).

4. Browne, 66.
5. Phillips, 53.
6. Darwin, *Journal of Researches*, 215.
7. Darwin, 216.
8. Darwin, 216.
9. Darwin, *The Origin of Species*, 65.
10. Darwin, 69.
11. Darwin, 419.
12. Darwin, 420.
13. Darwin, 420.
14. Darwin, 420.
15. Darwin, 426–27.
16. Phillips, 39.
17. Povinelli, 41–42.
18. Kohn, "Darwin's Ambiguities," 234.
19. Gould, *Full House*, 137.
20. Darwin, *Charles Darwin's Marginalia*, 164.
21. See, for example, Shanahan, *The Evolution of Darwinism*, 288.
22. Phillips, 39–40.
23. Phillips, 58.
24. Duncan, "We Were Never Human," 8–9.
25. Mayr, *The Growth of Biological Thought*, 335.
26. Duncan, *Human Forms*, 9.
27. T. Huxley, "*The Origin of Species*," 255.
28. T. Huxley, 278.
29. Phillips, 37.
30. Desmond, 501.
31. Levine, "Paradox," 84.
32. T. Huxley, *Evidence*, 132.
33. It is worth noting that Herbert Spencer would later take up this term in a decidedly non-ironic register in his theorization of "Transfigured Realism," which, as historian of science Chris Smith argues, is a form of "absolute idealism" traceable back to the metaphysics of early nineteenth-century German *naturphilosophie*. See "Darwin's Unsolved Problem," 119.

34. T. Huxley, 129–30.
35. Quoted in Desmond, 308.
36. Johnson, "The Pedestal and the Veil," 303.
37. K. Anderson, *Race and the Crisis of Humanism*, 147. Emphasis in original.
38. Hale, "Labor and the Human Relationship with Nature," 258.
39. Worster, *Nature's Economy*, 174.
40. T. Huxley, *Evolution and Ethics*, 10, 33.
41. T. Huxley, 19.

42. T. Huxley, 20.
43. T. Huxley, 83.
44. T. Huxley, 83.
45. T. Huxley, 44–45.
46. Darwin, *The Origin of Species*, 75, 73.
47. Grove, *Green Imperialism*, 484.
48. Stepan, *Picturing Tropical Nature*, 79.
49. Darwin, *The Variation of Animals and Plants*, 2.
50. Wallace, *The Malay Archipelago*, 445.
51. Wallace, *The Wonderful Century*, 190.
52. Ruskin, *The Storm Cloud*, 61.

53. To be clear, this is still not "climate change" as we know it today, though Wallace does at times sound like many late twentieth-century climate activists in his (understandable, if misguided) belief that increased public awareness was all that was needed for the culture to change course: "When this fact is thoroughly realized we shall surely put a stop to such a reckless and wholly unnecessary production of injurious smoke and dust" (85).

54. Wallace, *Darwinism*, 476.
55. Darwin, "To A. R. Wallace," 157.
56. "Geology: Its Progress," 365.
57. Ellegård, *Darwin and the General Reader*, 254.
58. Keith, *The Construction of Man's Family Tree*, 10.
59. Ellegård, 254.
60. See Bowler, *The Non-Darwinian Revolution*, 88.
61. Bowler, 184.
62. Stewart-Williams, *Darwin, God, and the Meaning of Life*, 64.
63. Gould, *Full House*, 19.
64. Gould, 19.
65. Petrinovich, *Darwinian Dominion*, 42.
66. Pauley, "Samuel Butler and His Darwinian Critics," 164.
67. Shaw, *Back to Methuselah*, 53.
68. Spencer, *First Principles*, 117.
69. Adorno and Horkheimer, *Dialectic of Enlightenment*, 43.
70. C. Taylor, 405.
71. Beer, *Darwin's Plots*, 39.
72. Tyndall, *Heat Considered*, 433.
73. Mivart, "Darwin's *Descent of Man*," 68–69.
74. Rylance, *Victorian Psychology*, 236.

75. Such arguments, it is worth noting, draw upon a long and deeply problematic history of philosophizing about the non-human animal as lacking self-conscious awareness. It is foundational to German *naturphilosophie*, for example, and the developmental visions of writers like Schelling, who saw natural history as the progressive realization of self-consciousness in human beings. A generation earlier, Immanuel Kant grants self-consciousness a similar pride of place, proposing a chillingly instrumental divide between humans and animals: "[They are] not self-conscious and are there merely as a means to

an end. That end is man" ("Duties towards Spirits and Animals," 239). Kant's moral philosophy, especially the idea of the categorical imperative, was foundational to Darwin's conception of moral evolution, which he works out in explicitly Lamarckian terms in the fourth chapter of *The Descent of Man*. For a discussion of Kant's influence on Darwin's moral philosophy, see Sideris, "One Step Up, Two Steps Back," 365–88.

76. Romanes, *Mental Evolution in Man*, 192.
77. Romanes, 212.
78. Romanes, 208.
79. Spencer, *The Principles of Psychology*, 1:147.
80. Spencer, 148.
81. Spencer, *The Principles of Ethics*, 63.
82. The idea of self-consciousness as a self-ratifying marker of essential difference persists well into the twentieth century and beyond, and from a wide array of political and philosophical positions. For example, in his influential *Introduction to the Reading of Hegel*, Alexander Kojève writes: "Man is Self-Consciousness. He is conscious of himself, conscious of his human reality and dignity; and it is in this that he is essentially different from animals" (3).
83. Khanna, *Dark Continents*, 1.
84. Spencer, *The Principles of Psychology*, 2:505.
85. Bagehot, *Physics and Politics*, 101.
86. Bagehot, 6.
87. Neyrat, "The Western Relation," 178.
88. Spencer, *Social Statics*, 33.
89. Rylance, 223.
90. Writing on Darwin and Comte in *From Comte to Benjamin Kidd*, the sociologist Robert Mackintosh uses decidedly Lamarckian terms about how the nineteenth century had witnessed the developmentalist trajectory of life on earth, "come to self-consciousness . . . and industrialism, the definitive social order, which corresponds to science or positivism, the definitive stage of thought, lies all around us, albeit still in sad confusion. The long regency of God is at an end. The minority of Humanity has ceased. . . . Mankind now at last may enter the land of promise" (21).
91. Darwin, *The Descent of Man*, 105.
92. Darwin, 105.
93. Knoll, "Dogs, Darwinism, and English Sensibilities," 15.
94. Darwin, 105.
95. Jackson, *Becoming Human*, 46.
96. Darwin, 193.
97. Darwin, 689.
98. Darwin, 689.
99. Fabian, *Time and the Other*, 33. Emphasis in original.
100. Darwin, *Journal of Researches*, 260.
101. Jackson, 35.
102. It has a pride of place in Hegel's thought, where it also functions to enforce a hierarchical division of humans from non-human animals, and Europeans from other

races: Africa, Hegel writes, "is the Unhistorical, Undeveloped Spirit, still involved in the conditions of mere nature" (*Philosophy of History*, 157.) The many points of continuity between certain influential strands of Victorian biology and Hegelian idealist evolutionism indicate the way that the valorization of self-consciousness, especially as it is connected to narratives of progress, has the status of what Bourdieu calls "doxa" in the nineteenth century and well beyond. See *Outline of a Theory of Practice*, 159.

103. Wald, "Science and Technology," 406.
104. Wald, 406.
105. J. Huxley, "Transhumanism," 13.
106. Wald, 408.
107. J. Huxley, "The Future of Man," 21–22.
108. Quoted in Zalloua, *Being Posthuman*, 66.
109. Zalloua, 67.
110. For Darwin's "formative" influence on Freud, see Ritvo, "The Impact of Darwin on Freud."
111. A. Freud, *The Ego*, 42.
112. S. Freud, *An Outline of Psycho-Analysis*, 61.
113. See Trunnell and Holt, "The Concept of Denial or Disavowal," 770.
114. Trunnell and Holt, 770.
115. See S. Freud, *Totem and Taboo*, 17, 98–99.
116. Swartz, *Psychoanalysis and Colonialism*, 17.
117. Nonini, "Freud, Anteriority and Imperialism," 27.
118. Nonini, 28, 31.
119. McClintock, *Imperial Leather*, 8.
120. S. Freud, *The Future of an Illusion* (trans. James Strachey), 21–22.
121. S. Freud, 43.
122. S. Freud, *The Future of an Illusion* (trans. Gregory Richter), 119.
123. S. Freud, *The Future of an Illusion* (trans. James Strachey), 5–6.
124. S. Freud, 22.
125. Allen, *Critique on the Couch*, 90.
126. See Deer, *Radical Animism*, 46–48. She is certainly right that Freud's attitude toward animism is far more complex than is usually allowed; the contradiction seems to have something to do with the official/unofficial position split discussed later.
127. Latour, *Facing Gaia*, 70.
128. S. Freud, *Civilization and Its Discontents*, 76–77.
129. S. Freud, 75–76.
130. S. Freud, 76.
131. S. Freud, *Introductory Lectures on Psycho-Analysis*, 284.
132. Peterson, *Monkey Trouble*, 15.
133. S. Freud, *Civilization and Its Discontents*, 103–4.
134. S. Freud, 104.
135. Moore and Fine, *Glossary*, 31.
136. Trunnel and Holt, 783.
137. Trunnel and Holt, 779.

138. Marcuse, *Eros and Civilization*, 11.
139. Whitebook, *Freud*, 146.
140. Whitebook, 162.
141. Whitebook, 350, 224.
142. S. Freud, *Introductory Lectures on Psycho-Analysis*, 280.
143. See Szabados, "Freud, Self-Knowledge and Psychoanalysis," 695–96.
144. S. Freud, *Totem and Taboo*, 258.
145. Zilcosky, "Savage Science," 468.
146. Zilcosky, 484.
147. Allen, 90.

Chapter Three

1. Plotz, "Speculative Naturalism," 53.
2. Plotz, 32.
3. Challenger, *How to Be Animal*, 120.
4. Gray, "The Mind's Body Problem."
5. Jameson, *The Political Unconscious*, 283–84.
6. Jameson, 283.
7. Stevens, *The Palm at the End of the Mind*, 382; Tennyson, *The Major Works*, 50.
8. Tennyson, 118.
9. Tennyson, 283.
10. Wells, *The Time Machine*, 77.
11. Tennyson, 283.
12. J. Miller, "Composing Decomposition," 389–90.
13. Gould, *Full House*, 20.
14. Tennyson, 282.
15. K. Anderson, 147.
16. Tennyson, 80.
17. Tennyson, 80.
18. Tennyson, 81.
19. The ambiguity in this final line has of course long been a subject of discussion in studies of Tennyson. John Peters discusses the both/and nature of this final exhortation, arguing that the speaker fashions himself as both leader and deceiver, "lead[ing] his mariners by seducing them as the Sirens had tried to seduce Ulysses in *The Odyssey*" ("To Strive, to Seek," 134). I find this reading convincing but would add that one of those fooled by this deception seems to be Ulysses himself. That deception is as much a mark *of* his aging as it is a response *to* it: no longer in command, he "edit[s] out the most intolerable part of his present reality," Arthur Ward argues, which is the frustration of his will ("'Ulysses' and 'Tithonus,'" 316). To return to the discussion of Arendt in the first chapter, we thus see in this figure something of her discussion of the modern, depoliticized notion of freedom as inward expression of the will in response to restricted circumstances.
20. Tucker, "The Fix of Form," 534.
21. Tennyson, 223.

22. Tucker, 534.
23. Tucker, 535.
24. Tennyson, 223–24.
25. Wells, 155–56. My emphasis.
26. Phillips, 29.
27. Wells, 156.
28. Yeats, *Collected Poems*, 41–42.
29. Williams, *Keywords*, 258.
30. Stevenson, *Dr. Jekyll and Mr. Hyde*, 79–80.
31. Saposnik, "The Anatomy of *Dr. Jekyll and Mr. Hyde*," 728.
32. Stevenson, 69, 93.
33. Stevenson, 82, 81, 85.
34. Stevenson, 83.
35. Stevenson, 90.
36. Stevenson, 90.

37. Critics of the novel have long focused on the complex pronominal play and the way it expresses the desire to enforce various forms of demarcation, as well as the impossibility of ever doing so. The classic reading is Peter Garrett's "Cries and Voices." See also Walker, *Labyrinths of Deceit*, 88.

38. Stevenson, 93.
39. Zhang, "Naming the Indescribable," 64.
40. Arata, "The Sedulous Ape," 253.

41. For example, Robert Mighall, in *A Geography of Victorian Gothic Fiction*, contextualizes Stevenson's novel in terms of Victorian evolutionary anthropology, which theorized "atavism" as the persistence of the "primitive" into the chronological present (139–53), linking it with lower-class criminality. As he puts it, "Jekyll not only internalizes the morality of his class, he internalizes its scientific models of hierarchy and deviance, the bestial and the primitive" (147). Arata opens up further complexities of the novel's treatment of atavism by showing how Hyde is used to suggest both upper- and lower-class forms of evolutionary degeneration, "how easily . . . [he] can be taken not for a brute but for a dandy" (237).

42. Stevenson, 84.
43. Conrad, *Heart of Darkness*, 14.
44. Conrad, 15.
45. Conrad, 14.
46. Conrad, 36.
47. J. Taylor, "Wilderness after Nature," 30.

48. For Caitlin Vandertop, Conrad's style in *Heart of Darkness* speaks to the idea that "the novella's irrealism and linguistic obscurity do not simply reveal the limits of subjective observation, but capture the real feeling of a region transformed by sociological abstractions to an unthinkable extent." "The Earth Seemed Unearthly," 693.

49. Vandertop, 693.
50. Conrad, 34.
51. Leavis, *The Great Tradition*, 219.

52. Worster, 174.
53. Achebe, "An Image of Africa," 788–79.
54. Conrad, 7.
55. Said, *Culture and Imperialism*, 20, 24.
56. Conrad, 76–77.
57. Dodds, *Psychoanalysis and Ecology at the Edge of Chaos*, 49.
58. Conrad, 7.
59. McConnell, *The Science Fiction of H. G. Wells*, 89.
60. Wells, *The Island of Dr. Moreau*, 131.
61. Wells, 131.
62. Wells, 131.
63. Wells, 8, 34, 52.
64. Wells, 60.
65. Wells, 73.
66. Wells, 75.
67. Bowler, 78–79.
68. Wells, 78.
69. Wells, 84.
70. Wells, 82.
71. Wells, 96.
72. Wells, 95.
73. Shakespeare, *King Lear*, 4.1.37–38.
74. Mbembe, "Necropolitics," 24.
75. Morton, *Ecology without Nature*, 118.
76. Wells, 131.
77. Wynter, 267.
78. Suvin, "Wells as the Turning Point," 110.
79. "R. Kipling: Comparative Psychologist," 858.
80. Kipling, *Writings on Writing*, 114.
81. See MacDuffie, "*The Jungle Books*."
82. Weismann, "All-Sufficiency," 309.
83. Kipling, *The Jungle Book*, 291.
84. As Jessica Straley argues in *Evolution and Imagination in Victorian Children's Literature*, *The Jungle Book* does not offer a straightforward linear trajectory for this process, as illustrated by the strange sequence here in which Mowgli moves from Ape to Buck to Man; nevertheless, the sense of heterogeneity and the various fanciful departures from Lamarckian orthodoxy are all premised upon and enabled by a strictly hierarchical scaffolding in which the final end point—"Man" assuming "his" rightful place as master—is never in doubt.

Chapter Four

1. Ghosh, 23.
2. Ghosh, 77.
3. Pascal, *The Dual Voice*, 19.

4. Eliot, *Middlemarch*, 620.
5. Eliot, 211.
6. Carroll, "*Middlemarch* and the Externality of Fact," 78–79.
7. Moretti, *Graphs, Maps, and Trees*, 87.
8. Eliot, 60.
9. Beer, *Darwin's Plots*, 147.
10. Kucich, "George Eliot and Objects," 321.
11. Duncan, *Human Forms*, 141.
12. Ferguson, "Jane Austen, *Emma*, and the Impact of Form," 159.
13. Moretti, 82.
14. See Pecora, *Self and Form in Modern Narrative*, especially the chapter on Joyce's "The Dead," 214–59.
15. D. Miller, *The Novel and the Police*, 25.
16. Ferguson, 172.
17. Eliot, *Daniel Deronda*, 19.
18. Joyce, *Dubliners*, 279.
19. Trollope, *The Last Chronicle of Barset*, 127.
20. Pascal, 80.
21. D. Miller, *Jane Austen*, 71.
22. Eliot, *Middlemarch*, 144.
23. Freedgood, *Worlds Enough*, 70–71.
24. Chaitin, "Listening Power," 1023.
25. Flaubert, *Madame Bovary*, 68.
26. Hardy, *Jude the Obscure*, 122–23.
27. Tolstoy, *The Death of Ivan Ilyich*, 67–68.
28. Pascal, 78.
29. James, *Washington Square*, 11.
30. D. Miller, 71.
31. Eliot, *The Mill on the Floss*, 481.
32. Pascal, 80.
33. Freedgood, "The Novelist and Her Poor," 219.
34. Finch and Bowen, "The Tittle-Tattle of Highbury," 5; Freedgood, *Worlds Enough*, 71.
35. Ferguson, "Planetary Literary History," 676.
36. Hardy, 85–86.
37. Eliot, *Daniel Deronda*, 380.
38. Cohen, 22.
39. Eliot, *Middlemarch*, 529.
40. Eliot, 10.
41. Oldfield, "The Language of the Novel," 69.
42. Eliot, 274.
43. Eliot, 10.
44. Eliot, 19.
45. Eliot, 51.

Notes to Chapter 4

46. Eliot, 37.
47. Eliot, 44.
48. Pascal, 86.
49. Pascal, 86.
50. Eliot, 525.
51. Cohen, 7–8.
52. Eliot, 687.
53. Miller, Hardy, and Poirier, "*Middlemarch*, Chapter 85," 444–45.
54. Wood, *How Fiction Works*, 9. I'd add briefly here, in a note of appreciation, that Wood's classes on free indirect discourse in novels by V. S. Naipaul, Henry Green, Saul Bellow, and others were foundational to my own thinking about the technique.
55. Eliot, 824.
56. Eliot, 824.
57. Miller, Hardy, and Poirier, 444.
58. Eliot, 478.
59. Eliot, 523.
60. Eliot, 620.
61. Carroll, "*Middlemarch* and the Externality of Fact," 84.
62. Eliot, 95.
63. Levine, "George Eliot's Hypothesis of Reality," 1.
64. Eliot, *Life and Letters*, 549.
65. Eliot, *Middlemarch*, 211.
66. Levine, 14.
67. Eliot, 795–96.
68. Eliot, 796.
69. Gallagher, "George Eliot," 72.
70. Eliot, 194.
71. Levine, "Sounds of Silence," 146.
72. Kreilkamp, *Minor Creatures*, 167.
73. Eliot, *The Mill on the Floss*, 251.
74. Eliot, 251.
75. Plumwood, 21.
76. T. Huxley, *On the Physical Basis of Life*, 8.
77. Eliot, *Miscellaneous Essays*, 247.
78. J. Miller, "The Ecological Plot," 174.
79. Darwin, *The Descent of Man*, 149.
80. Darwin, 147.
81. Duncan, *Human Forms*, 192.
82. Darwin, 147.
83. Jackson, 55.
84. Miller, Hardy, and Poirier, 447–48.
85. J. Miller, 175.
86. Kingstone, "Human-Animal Elision," 90.
87. Eliot, *Middlemarch*, 211.

88. Eliot, 838.
89. Chakrabarty, 201.
90. See Michael Tondre's superb essay "George Eliot's 'Fine Excess'" on the way Eliot uses a thermodynamic vocabulary in this passage and elsewhere in *Middlemarch* to discuss the moral life; as he puts it, "the term 'diffusion' expressed a set of newly perceived phenomena about vital energy" (204–5).
91. Darwin, *The Origin of Species*, 425.
92. Wallace, "On the Origin of the Human Races," clxviii.
93. Poster, *Cultural History and Postmodernity*, 59.
94. E. Thompson, "Socialist Humanism," 122.
95. Kingstone, 101.
96. Eliot, *Miscellaneous Essays*, 402.
97. Eliot, *Daniel Deronda*, 124.
98. Eliot, 53.
99. Eliot, 339.
100. E. Miller, *Extraction Ecologies*, 27.
101. Eliot, 593–94.
102. James, "*Daniel Deronda*: A Conversation," 73.
103. Duncan, *Human Forms*, 191.
104. Eliot, 90.
105. Eliot, 431.
106. Eliot, 93.
107. Eliot, 140.
108. Eliot, 757.
109. Eliot, 93, 280.
110. Duncan, 168.
111. Eliot, 63.
112. Duncan, 190.
113. Eliot, 63–64.
114. Eliot, 40.
115. Eliot, 277.
116. Penner, "Unmapped Country," 92, 91.
117. Stone, "George Eliot's *Daniel Deronda*," 58.
118. Eliot, 141.
119. Eliot, 143.
120. Eliot, 292.
121. Eliot, 328.
122. Cohen, *States of Denial*, 7.
123. Eliot, 146.
124. Eliot, 376–77.
125. Eliot, 276.
126. Eliot, 137.
127. Leavis, 106.
128. Eliot, 303.

129. Eliot, 111.
130. Eliot, 696.
131. Eliot, 762.
132. Eliot, 804.
133. Eliot, 520.
134. Eliot, 379.
135. Eliot, 380.
136. Eliot, 380.
137. Eliot, 380.
138. Eliot, 381.
139. Leavis, 84–85.
140. A. Anderson, *The Powers of Distance*, 121.
141. A. Anderson, 145–46.
142. Stević, "Convenient Cosmopolitanism," 593.
143. A. Anderson, 133.
144. Stević, 607.
145. Stević, 600.
146. Duncan, "George Eliot's Science Fiction," 29–30.

Chapter Five

1. Woolf, *On Being Ill*, 39.
2. See, for example, Gillian Beer's *Open Fields*, especially chapter 10; and MacDuffie, *Victorian Literature*, especially chapter 2.
3. Woolf, 38–39.
4. Woolf, 35.
5. Woolf, 36.
6. Woolf, 36–37.
7. Conrad, 186.
8. Woolf, *The Common Reader*, 150.
9. Latour, *Facing Gaia*, 10.
10. Morton, *The Ecological Thought*, 8.
11. Beer, "Virginia Woolf and Pre-History," 100. Bonnie Scott's *In the Hollow of the Wave*, building on Beer's work, extensively describes Woolf's interest in the findings from a wide array of scientific disciplines, especially Darwinian biology, and reveals the complex ways she challenged many of the implicit and explicit patriarchal biases in the work of Darwin and other evolutionary thinkers.
12. Woolf, *Between the Acts*, 9.
13. Woolf, 9.
14. Woolf, *Mrs. Dalloway*, 77.
15. Woolf, 12.
16. Tromanhauser, "Mrs. Dalloway's 'Animals,'" 191, 190.
17. Woolf, 13.
18. K. Anderson, 149.
19. Woolf, 28.

20. Manya Lempert aligns Woolf with Darwin through the former's sense of "tragic form," which "embodies an ethics of human limitation and independence." Lempert argues that "apart from Darwin's account of evolution as a process unfolding without design, nineteenth- and twentieth-century interpretations of natural history continued to cast the saga of human origins as a divine comedy whose *telos* was man" ("Virginia Woolf, Charles Darwin, and the Rebirth of Tragedy," 449–50). In some ways this disconnect is part of the denialism I'm discussing here, though I would argue that this sense of telos is deeply embedded even within overtly "Darwinian" scientific discourse and, at times, even in Darwin's own writings.

21. Woolf, 28.
22. Woolf, *Jacob's Room*, 6.
23. Woolf, 7.
24. Though it's perhaps worth mentioning that the non-human images in *Jacob's Room* do seem to function more clearly as symbolic commentaries on the human plot: the view from the grass and the Sisyphean motions of the crab look forward to Jacob's death and the futility of the Great War, which will obviously loom large at the end of the novel.
25. Woolf, *The Diary of Virginia Woolf*, 2:13.
26. Darwin, *The Origin of Species*, 74.
27. James, "Preface to *Roderick Hudson*," 5.
28. James, 6.
29. Stevens, 98.
30. Woolf, *To the Lighthouse*, 16.
31. Woolf, 105–6.
32. Woolf, *The Voyage Out*, 207–8.
33. Woolf, 208.
34. Woolf, *To the Lighthouse*, 35–36.
35. Horton, "The Trans-scalar Challenge of Ecology," 8.
36. T. Huxley, *Evidence*, 132.
37. T. Clark, 74.
38. Woolf, 38.
39. Woolf, 38.
40. Woolf, 39.
41. Woolf, 104–5.
42. Woolf, 100.
43. Woolf, 97.
44. Woolf, 105.
45. Woolf, 102–3.
46. Plumwood, 21.
47. Woolf, 38.
48. Woolf, 37–38.
49. Woolf, 128.
50. Woolf, 131.
51. Woolf, 180.
52. Plumwood, 21.

53. Woolf, 132, 129.
54. Dickens, *Martin Chuzzlewit*, 7.
55. J. Taylor, *The Sky of Our Manufacture*, 5.
56. Dickens, *Our Mutual Friend*, 149.
57. Dickens, 150.
58. Woolf, 133.
59. Woolf, 133.
60. Woolf, 127–29.
61. Woolf, *The Letters of Virginia Woolf*, 280.
62. Levenback, *Virginia Woolf and the Great War*, 11.
63. Woolf, *The Diary of Virginia Woolf*, 1:32.
64. Woolf, 131.
65. "Treasure Houses," 199.
66. Outka, "The Transitory Space," 65.
67. H. Smith, *Not So Quiet*, 30–31.
68. Woolf, *The Letters of Virginia Woolf*, 76.
69. T. Clark, 159; Woolf, *The Diary of Virginia Woolf*, 1:65.
70. Latour, *Facing Gaia*, 10.
71. C. Clark, *The Sleepwalkers*, xxix.
72. Morton, 17.
73. Deese, *Climate Change*, 19.
74. Clifford, 156.
75. H. Scott, "Industrial Souls," 591.

Conclusion

1. Atwood, *The Year of the Flood*, 239.
2. See Hengen, "Metaphysical Romance," 154–56.
3. "Over the five years of high school that were mandated there and then, we studied two novels each of George Eliot, Charles Dickens, and Thomas Hardy, and we studied them very thoroughly." Atwood, *In Other Worlds*, 40. For her interest in *Wuthering Heights*, see 143.
4. Snell, "Margaret Atwood."
5. Atwood, 156.
6. Chakrabarty, 220.
7. Adorno and Horkheimer, 95.
8. Atwood, *Oryx and Crake*, 253–54.
9. Atwood, 184.
10. Dickens, *David Copperfield*, 13.
11. Povinelli, 39.
12. Povinelli, 59.
13. Atwood, 3.
14. Atwood, 119–20.
15. Dostoevsky, 261.
16. Atwood, *The Year of the Flood*, 51.

17. Atwood, *Oryx and Crake*, 167.
18. Atwood, 167–68.
19. Atwood, 166.
20. Atwood, 166.
21. Atwood, *The Year of the Flood*, 241.
22. Atwood, *Oryx and Crake*, 53.
23. Atwood, *MaddAddam*, 317.
24. Atwood, 333.
25. Atwood, *The Year of the Flood*, 33.
26. Duncan, *Human Forms*, 3.
27. Atwood, *Oryx and Crake*, 7–8.
28. The original quote is from Book IV, chapter XII (page 245 of the Norton Critical Edition). I've quoted here from an epigraph in Atwood's novel because she modernizes the pronouns, punctuates differently, and adds a word not in the original.
29. Woolf, *To the Lighthouse*, 147.
30. Jameson, "Review: *Of Islands and Trenches*," 11.
31. Atwood, *Oryx and Crake*, 69.
32. C. Taylor, 417.
33. Atwood, 57.
34. Atwood, 57.
35. Kucich, 321.
36. Mbembe, *Out of the Dark Night*, 22.
37. Atwood, *MaddAddam*, 97.
38. Atwood, 91.
39. Atwood, 150.
40. Atwood, *The Year of the Flood*, 16.
41. Atwood, 33.
42. Atwood, 170.
43. Atwood, *MaddAddam*, 154.
44. Atwood, *The Year of the Flood*, 248.
45. Atwood, 97.
46. Atwood, *MaddAddam*, 111.
47. Fabre, *The Hunting Wasps*, 192.
48. Darwin, "To J-H Fabre," 52.
49. Dillard, 64.
50. See Cheney's "The Waters of Separation" for a brief but intriguing comment on the connection between Atwood and Dillard on the question of whether the form of a text functions "to block perception" (51).
51. Darwin, 52.
52. Fabre, *Souvenirs entomologiques*, 323. "Alas! you have left for a better place, not yet knowing anything of the book beyond its first few lines! May your name at least appear there, carried by some of those industrious and beautiful Hymenoptera that you loved so much" (my translation, with an assist from my father-in-law).
53. Darwin, 52.

54. Darwin, "To Asa Gray," 224.
55. "It may be difficult, but we ought to admire the savage instinctive hatred of the queen-bee, which urges her instantly to destroy the young queens her daughters as soon as born, or to perish herself in the combat; for undoubtedly this is for the good of the community; and maternal love or maternal hatred, though the latter fortunately is most rare, is all the same to the inexorable principle of natural selection." Darwin, *The Origin of Species*, 184–85.
56. "We hear from mathematicians that bees have practically solved a recondite problem, and have made their cells of the proper shape to hold the greatest possible amount of honey, with the least possible consumption of precious wax in their construction. It has been remarked that a skilful workman, with fitting tools and measures, would find it very difficult to make cells of wax of the true form, though this is perfectly effected by a crowd of bees working in a dark hive. Grant whatever instincts you please, and it seems at first quite inconceivable how they can make all the necessary angles and planes, or even perceive when they are correctly made. But the difficulty is not nearly so great as it at first appears: all this beautiful work can be shown, I think, to follow from a few very simple instincts." Darwin, 203–4.
57. Darwin, 74.
58. Darwin, 74–75.
59. Haraway, *The Companion Species Manifesto*, 15.
60. Hustak and Myers, "Involuntary Momentum," 77, 79.
61. Carson, *Silent Spring*, 73.
62. Mates, Perfecto, and Badgley, "Parasitoid Wasp Diversity," 82.
63. Dillard, 170.
64. Dillard, 170–71.

BIBLIOGRAPHY

Achebe, Chinua. "An Image of Africa." *Massachusetts Review* 18.4 (1977): 782–94.
Adorno, Theodor, and Max Horkheimer. *Dialectic of Enlightenment: Philosophical Fragments*. Ed. Gunzelin Schmid Noerr. Trans. Edmund Jephcott. 1947. Reprint, Stanford University Press, 2002.
Allen, Amy. *Critique on the Couch: Why Critical Theory Needs Psychoanalysis*. Columbia University Press, 2021.
Anderson, Amanda. *The Powers of Distance: Cosmopolitanism and the Cultivation of Detachment*. Princeton University Press, 2001.
Anderson, Kay. *Race and the Crisis of Humanism*. Routledge, 2007.
Arata, Stephen D. "The Sedulous Ape: Atavism, Professionalism, and Stevenson's *Jekyll and Hyde*." *Criticism* 37.2 (1995): 233–59.
Arendt, Hannah. "What Is Freedom?" In *Between Past and Future*, 143–71. Penguin, 1961.
Atwood, Margaret. *In Other Worlds: SF and the Human Imagination*. Doubleday, 2011.
———. *MaddAddam*. Anchor, 2013.
———. *Oryx and Crake*. Anchor, 2003.
———. *The Year of the Flood*. Anchor, 2009.
Bagehot, Walter. *Physics and Politics: Or, Thoughts on the Application of the Principles of Natural Selection and Inheritance to Political Society*. D. Appleton, 1873.
Barrish, Phillip. "Accumulating Variation: Darwin's *On the Origin of Species* and Contemporary Literary and Cultural Theory." *Victorian Studies* 34.4 (1991): 431–53.
Bateson, Gregory. *Steps to an Ecology of Mind*. University of Chicago Press, 1972.
Becker, Ernest. *Escape from Evil*. Free Press, 1975.
Beer, Gillian. *Darwin's Plots: Evolutionary Narrative in Darwin, George Eliot, and Nineteenth-Century Fiction*. 1983. Reprint, Cambridge University Press, 2000.
———. *Open Fields: Science in Cultural Encounter*. Clarendon Press, 1996.

———. "Virginia Woolf and Pre-History." In *Virginia Woolf: A Centenary Perspective*, ed. Eric Warner, 99–123. Palgrave, 1984.
Benjamin, Walter. *The Arcades Project*. Trans. Howard Eiland and Kevin McLaughlin. Harvard University Press, 1999.
———. *One-Way Street and Other Writings*. Trans. Edmund Jephcott and Kingsley Shorter. Harcourt, 1979.
Berlant, Lauren. *Cruel Optimism*. Duke University Press, 2011.
Bilgrami, Akeel. *Secularism, Identity, and Enchantment*. Harvard University Press, 2014.
Bouissac, Paul. "What Is a Human? Ecological Semiotics and the New Animism." *Semiotica* 77.4 (1989): 497–516.
Bourdieu, Pierre. *Outline of a Theory of Practice*. Cambridge University Press, 1977.
Bowler, Peter. *The Non-Darwinian Revolution: Reinterpreting a Historical Myth*. Johns Hopkins University Press, 1988.
Brontë, Charlotte. "Letter to James Taylor," 11 February 1851. In *The Letters of Charlotte Brontë 1848–1851*, ed. Margaret Smith, 564–65. Oxford University Press, 1995.
Browne, Janet. *Charles Darwin: The Power of Place*. Vol. 2. Princeton University Press, 2003.
Carroll, David. "*Middlemarch* and the Externality of Fact." In *This Particular Web: Essays on* Middlemarch, ed. Ian Adam, 73–90. University of Toronto Press, 1975.
Carson, Rachel. *Lost Woods: The Discovered Writing of Rachel Carson*. Beacon Press, 2011.
———. *Silent Spring*. 1962. Houghton Mifflin, 1994.
Chaitin, Gilbert D. "Listening Power: Flaubert, Zola, and the Politics of *Style Indirect Libre*." *French Review* 72.6 (1999): 1023–37.
Chakrabarty, Dipesh. "The Climate of History: Four Theses." *Critical Inquiry* 35.2 (2009): 197–222.
Challenger, Melanie. *How to Be Animal: A New History of What It Means to Be Human*. Penguin, 2021.
Cheney, Jim. "'The Waters of Separation': Myth and Ritual in Annie Dillard's *Pilgrim at Tinker Creek*." *Journal of Feminist Studies in Religion* 6.1 (1990): 41–63.
Clark, Christopher. *The Sleepwalkers: How Europe Went to War in 1914*. HarperCollins, 2012.
Clark, Timothy. *Ecocriticism on the Edge: The Anthropocene as a Threshold Concept*. Bloomsbury, 2015.
Clifford, William Kingdon. *Seeing and Thinking*. 2nd ed. Macmillan, 1880.
Cohen, Stanley. *States of Denial: Knowing about Atrocities and Suffering*. Wiley, 2001.
Coleridge, Samuel Taylor. *Complete Works*. Vol. 6. Harper, 1868.
Conrad, Joseph. *Heart of Darkness*. 1899. Reprint, Norton Critical Edition, 1988.
"Darwin and *The Descent of Man*." *Edinburgh Review* 134 (July 1871): 195–235.
Darwin, Charles. *Charles Darwin's Marginalia*. Ed. Mario DiGregorio. Garland, 1990.
———. *The Descent of Man*. Ed. James Moore and Adrian Desmond. 1871. Reprint, Penguin, 2004.
———. *Journal of Researches*. 1839. Reprint, John Murray, 1889.
———. *On the Origin of Species by Means of Natural Selection*. Ed. William Bynum. 1859. Reprint, Penguin, 2009.

———. "To A. R. Wallace," March 27, 1869. In *The Correspondence of Charles Darwin*, vol. 17, ed. Frederick Burkhardt and James Secord, 156–57. Cambridge University Press, 2009.

———. "To Asa Gray," May 22, 1860. In *The Correspondence of Charles Darwin*, vol. 8, ed. Frederick Burkhardt, Duncan Porter, Janet Browne, and Marsha Richmond, 223–24. Cambridge University Press, 1993.

———. "To Emma Darwin," April 28, 1858. In *The Correspondence of Charles Darwin*, vol. 7, ed. Frederick Burkhardt and Sydney Smith, 84–85. Cambridge University Press, 1991.

———. "To J. D. Hooker," April 7, 1859. In *The Correspondence of Charles Darwin*, vol. 7, ed. Frederick Burkhardt and Sydney Smith, 280. Cambridge University Press, 1991.

———. "To J. D. Hooker," May 31, 1866. In *The Correspondence of Charles Darwin*, vol. 14, ed. Frederick Burkhardt and Duncan M. Porter, 194. Cambridge University Press, 2004.

———. "To J-H Fabre," January 31, 1880. In *The Correspondence of Charles Darwin*, vol. 28, ed. Frederick Burkhardt and James Secord, 51–53. Cambridge University Press, 2021.

———. *The Variation of Animals and Plants under Domestication*. 1868. Reprint, D. Appleton, 1896.

Deer, Jemma. *Radical Animism: Reading for the End of the World*. Bloomsbury, 2020.

Deese, R. S. *Climate Change and the Future of Democracy*. Springer, 2019.

Desmond, Adrian. *Huxley: From Devil's Disciple to Evolution's High Priest*. Basic Books, 1997.

Desmond, Adrian, and James Moore. *Darwin: The Life of a Tormented Evolutionist*. Norton, 1991.

Dickens, Charles. *David Copperfield*. Ed. Jeremy Tambling. 1850. Reprint, Penguin, 2004.

———. *Martin Chuzzlewit*. 1844. Reprint, J. R. Osgood, 1872.

———. *Our Mutual Friend*. 1870. Reprint, Chapman & Hall, 1880.

Dickinson, Janis L. "The People Paradox: Self-Esteem Striving, Immortality Ideologies, and Human Response to Climate Change." *Ecology and Society* 14.1 (2009). http://www.jstor.com/stable/26268058.

Didion, Joan. *The White Album*. 1979. Reprint, Macmillan, 1990.

Dillard, Annie. *Pilgrim at Tinker Creek*. 1974. Reprint, Harper Perennial, 1988.

Dodds, Joseph. *Psychoanalysis and Ecology at the Edge of Chaos: Complexity Theory, Deleuze, Guattari, and Psychoanalysis for a Climate in Crisis*. Routledge, 2012.

Dostoevsky, Fyodor. *The Brothers Karamazov*. Trans. Richard Pevear and Larissa Volokhonsky. 1880. Reprint, North Point Press, 1990.

Duncan, Ian. "George Eliot's Science Fiction." *Representations* 125.1 (2014): 15–39.

———. *Human Forms: The Novel in the Age of Evolution*. Princeton University Press, 2019.

———. "We Were Never Human: Monstrous Forms of Nineteenth-Century Fiction." In *Victorian Transformations: Genre, Nationalism and Desire in Nineteenth-Century Literature*, ed. Bianca Tredennick, 7–27. Routledge, 2011.

Eliot, George. *Daniel Deronda*. Ed. Terence Cave. 1876. Reprint, Penguin, 1995.

———. *Life and Letters*. Vol. 2. G. Sproul, 1899.
———. *The Lifted Veil and Brother Jacob*. Tauchnitz, 1878.
———. *Middlemarch*. Ed. Rosemary Ashton. 1872. Reprint, Penguin, 1994.
———. *The Mill on the Floss*. Ed. Gordon Haight. 1860. Reprint, Oxford University Press, 1999.
———. *Miscellaneous Essays: Impressions of Theophrastus Such, The Veil Lifted, Brother Jacob*. Doubleday, 1901.
Ellegård, Alvar. *Darwin and the General Reader: The Reception of Darwin's Theory of Evolution in the British Periodical Press, 1859–1872*. University of Chicago Press, 1990.
Fabian, Johannes. *Time and the Other: How Anthropology Makes Its Object*. Columbia University Press, 1983.
Fabre, Jean-Henri. *The Hunting Wasps*. Trans. Alexander Texiera de Mattos. Dodd, Mead, 1915.
———. *Souvenirs entomologiques: Études sur l'instinct et les moeurs des insectes*. C. Delagrave, 1879.
Fanon, Frantz. *The Wretched of the Earth*. Trans. Constance Farrington. Grove Press, 1963.
Ferguson, Frances. "Jane Austen, *Emma*, and the Impact of Form." *MLQ: Modern Language Quarterly* 61.1 (2000): 157–80.
———. "Planetary Literary History: The Place of the Text." *New Literary History* 39.3 (2008): 657–84.
Finch, Casey, and Peter Bowen. "The Tittle-Tattle of Highbury: Gossip and the Free Indirect Style in *Emma*." *Representations* 31 (1990): 1–18.
Flaubert, Gustave. *Madame Bovary*. Trans. Francis Steegmuller. 1856. Reprint, Knopf, 1991.
Fowles, John. *The French Lieutenant's Woman*. 1969. Reprint, Little, Brown, 1998.
Freedgood, Elaine. "The Novelist and Her Poor." *Novel: A Forum on Fiction* 47.2 (2014): 210–23.
———. *Worlds Enough: The Invention of Realism in the Victorian Novel*. Princeton University Press, 2019.
Freud, Anna. *The Ego and the Mechanisms of Defense*. International Universities Press, 1966.
Freud, Sigmund. *Civilization and Its Discontents*. Ed. and trans. James Strachey. Norton, 2005.
———. *The Future of an Illusion*. Ed. and trans. James Strachey. Norton, 1961.
———. *The Future of an Illusion*. Ed. Todd Dufresne. Trans. Gregory Richter. Broadview, 2012.
———. *Introductory Lectures on Psycho-Analysis*. Ed. and trans. James Strachey. In *The Standard Edition of the Complete Works of Sigmund Freud*, vol. 16. Hogarth Press, 1963.
———. *An Outline of Psycho-Analysis*. Ed and trans. James Strachey. Norton, 1969.
———. *Totem and Taboo*. Trans. A. A. Brill. Moffat, Yard, 1918.
Gallagher, Catherine. "George Eliot: Immanent Victorian." *Representations* 90.1 (2005): 61–74.

Garrett, Peter. "Cries and Voices: Reading *Jekyll and Hyde*." In *Dr. Jekyll and Mr. Hyde after One Hundred Years*, ed. William Veeder and Gordon Hirsch, 59–72. University of Chicago Press, 1988.
"Geology and Protestantism." *Dublin Review* 44 (June 1858): 375–95.
"Geology: Its Progress and Limits as a Science." *British and Foreign Evangelical Review* 15 (1866): 353–72.
Ghosh, Amitav. *The Great Derangement: Climate Change and the Unthinkable*. University of Chicago Press, 2016.
Glasson, Ben. "Aware but Unmoved: Symptoms of the Absent Centre of Climate Change Discourse." Paper presented to the Australian Political Studies Annual Conference, Hobart, Tasmania, September 2012.
Gosse, Phillip. *Omphalos: An Attempt to Untie the Geological Knot*. J. Van Voorst, 1857.
Gould, Stephen Jay. *Ever Since Darwin: Reflections in Natural History*. Norton, 1992.
———. *Full House: The Spread of Excellence from Plato to Darwin*. Harvard University Press, 2011.
Gray, John. "The Mind's Body Problem." *New York Review of Books*, December 2, 2021. https://www.nybooks.com/articles/2021/12/02/the-minds-body-problem/.
Griffiths, Devin. "The Ecology of Form." *Critical Inquiry* 48.1 (2021): 68–93.
Grosz, Elizabeth. *Becoming Undone: Darwinian Reflections on Life, Politics and Art*. Duke University Press, 2011.
Grove, Richard. *Green Imperialism: Colonial Expansion, Tropical Island Edens and the Origins of Environmentalism, 1600–1860*. Cambridge University Press, 1996.
Hale, Piers. "Labor and the Human Relationship with Nature: The Naturalization of Politics in the Work of Thomas Henry Huxley, Herbert George Wells, and William Morris." *Journal of the History of Biology* 36.2 (2003): 249–84.
Hall, Catherine, and Daniel Pick. "Thinking about Denial." *History Workshop Journal* 84 (2017): 1–23.
Hansson, Sven Ove. "Science Denial as a Form of Pseudoscience." *Studies in History and Philosophy of Science* 63 (2017): 39–47.
Haraway, Donna Jeanne. *The Companion Species Manifesto: Dogs, People, and Significant Otherness*. Prickly Paradigm Press, 2003.
Hardy, Thomas. *Jude the Obscure*. Ed. Patricia Ingham. 1895. Reprint, Oxford World's Classics, 2002.
Hegel, Georg Wilhelm Friedrich. *Philosophy of History*. Trans. John Sibree. P. F. Collier, 1901.
Hengen, Shannon. "'Metaphysical Romance': Atwood's Ph.D. Thesis and *The Handmaid's Tale*." *Science Fiction Studies* 18.1 (1991): 154–56.
Herbert, Christopher. *Victorian Relativity: Radical Thought and Scientific Discovery*. University of Chicago Press, 2001.
Horton, Zach. "The Trans-scalar Challenge of Ecology." *ISLE: Interdisciplinary Studies in Literature and Environment* 26.1 (2019): 5–26.
Hustak, Carla, and Natasha Myers. "Involuntary Momentum: Affective Ecologies and the Science of Plant/Insect Encounters." *differences: A Journal of Feminist Cultural Studies* 15.3 (2012): 74–118.

Huxley, Julian. "The Future of Man: Evolutionary Aspects." In *Man and His Future*, ed. Gordon Wolstenholme, 1–22. Little, Brown, 1963.

———. "Transhumanism." In *New Bottles for New Wine*, 13–17. Chatto & Windus, 1957.

Huxley, Thomas Henry. *Evidence as to Man's Place in Nature*. D. Appleton, 1863.

———. *Evolution and Ethics and Other Essays*. Macmillan, 1894.

———. "Geological Reform." In *Lay Sermons, Addresses and Reviews*, 228–54. D. Appleton, 1874.

———. *On the Physical Basis of Life*. College Courant, 1869.

———. "*The Origin of Species*." In *Lay Sermons, Addresses and Reviews*, 255–98. D. Appleton, 1874.

Jackson, Zakiyyah Iman. *Becoming Human: Matter and Meaning in an Antiblack World*. New York University Press, 2020.

James, Henry. "*Daniel Deronda*: A Conversation." In *Partial Portraits*, 65–93. Macmillan, 1888.

———. "Preface to *Roderick Hudson*." In *The Art of the Novel: Critical Prefaces*, 3–19. University of Chicago Press, 2011.

———. *Washington Square*. Ed. Philip Horne and Martha Banta. 1880. Reprint, Penguin, 2007.

Jameson, Fredric. *The Antinomies of Realism*. Verso, 2013.

———. *The Political Unconscious: Narrative as a Socially Symbolic Act*. 1981. Reprint, Cornell University Press, 2015.

———. "Review: *Of Islands and Trenches: Naturalization and the Production of Utopian Discourse*." *Diacritics* 7.2 (1977): 2–21.

"Jefferey's British Conchology: Snails." *Blackwood's Edinburgh Magazine* 92 (1862): 77–91.

Johnson, Walter. "The Pedestal and the Veil: Rethinking the Capitalism/Slavery Question." *Journal of the Early Republic* 24.2 (2004): 299–308.

Joyce, James. *Dubliners*. B. W. Huebsch, 1925.

Kant, Immanuel. "Duties towards Spirits and Animals." In *Lectures on Ethics*, trans. Louis Infield, 239–41. Century, 1930.

Karera, Axelle. "Blackness and the Pitfalls of Anthropocene Ethics." *Critical Philosophy of Race* 7.1 (2019): 32–56.

Keith, Arthur. *The Construction of Man's Family Tree*. Watts, 1934.

Khanna, Ranjana. *Dark Continents: Psychoanalysis and Colonialism*. Duke University Press, 2003.

Kingstone, Elizabeth. "Human-Animal Elision: A Darwinian Universe in George Eliot's Novels." *Nineteenth-Century Contexts* 40.1 (2018): 87–103.

Kipling, Rudyard. *The Jungle Book*. Ed. W. W. Robson. 1894–95. Reprint, Oxford University Press, 2008.

———. *Writings on Writing*. Ed. Sandra Kemp and Lisa Lewis. Cambridge University Press, 1996.

Klein, Naomi. *This Changes Everything: Capitalism versus the Climate*. Simon & Schuster, 2014.

Knoll, Elizabeth. "Dogs, Darwinism, and English Sensibilities." In *Anthropomorphism, Anecdotes, and Animals*, ed. Robert W. Mitchell, 12–21. SUNY Press, 1997.

Kohn, David. "Darwin's Ambiguity: The Secularization of Biological Meaning." *British Journal for the History of Science* 22.2 (1989): 215–39.

Kojève, Alexander. *Introduction to the Reading of Hegel*. Ed. Allan Bloom. Trans. James H. Nichols. Cornell University Press, 1980.

Kreilkamp, Ivan. *Minor Creatures: Persons, Animals, and the Victorian Novel*. University of Chicago Press, 2018.

Kucich, John. "George Eliot and Objects: Meaning as Matter in *The Mill on the Floss*." *Dickens Studies Annual* 12 (1983): 319–40.

Latour, Bruno. *Facing Gaia: Eight Lectures on the New Climatic Regime*. John Wiley & Sons, 2017.

———. "Will Non-humans Be Saved? An Argument in Ecotheology." *Journal of the Royal Anthropological Institute* 15.3 (2009): 459–75.

Leavis, F. R. *The Great Tradition*. Doubleday, 1954.

Lempert, Manya. "Virginia Woolf, Charles Darwin, and the Rebirth of Tragedy." *Twentieth Century Literature* 64.4 (2018): 449–82.

Levenback, Karen. *Virginia Woolf and the Great War*. Syracuse University Press, 1999.

Levine, George. *Darwin Loves You: Natural Selection and the Re-enchantment of the World*. Princeton University Press, 2008.

———. "George Eliot's Hypothesis of Reality." *Nineteenth-Century Fiction* 35.1 (1980): 1–28.

———. "Paradox: The Art of Scientific Naturalism." In *Victorian Scientific Naturalism: Community, Identity, Continuity*, ed. Bernard Lightman and Gowan Dawson, 79–100. University of Chicago Press, 2014.

———. "The Sounds of Silence." *Raritan* 40.1 (2020): 141–63.

Macdonald, Nancy. "Naomi Klein on Climate Change and Capitalism." *Maclean's*, September 13, 2014. https://macleans.ca/society/the-interview-climate-activist-naomi-klein/.

MacDuffie, Allen. "Charles Darwin and the Victorian Pre-History of Climate Denial." *Victorian Studies* 16.4. (2018): 543–64.

———. "*The Jungle Books*: Rudyard Kipling's Lamarckian Fantasy." *PMLA* 129.1 (2014): 18–34.

———. *Victorian Literature, Energy, and the Ecological Imagination*. Cambridge University Press, 2014.

Mackintosh, Robert. *From Comte to Benjamin Kidd: The Appeal to Biology or Evolution for Human Guidance*. Macmillan, 1899.

Malm, Andreas. *White Skin, Black Fuel: On the Danger of Fossil Fascism*. Verso, 2021.

Mannoni, Octave. "I Know Well, but All the Same." Trans. G. M. Goshgarian. In *Perversion and the Social Relation*, ed. Molly Anne Rothenberg, Dennis Foster, and Slavoj Žižek, 68–92. Duke University Press, 2003.

Marcuse, Herbert. *Eros and Civilization: A Philosophical Inquiry into Freud*. 1966. Reprint, Beacon Press, 2015.

Marx, Karl. *Capital: A Critical Analysis of Capitalist Production*. 3rd German ed. Trans. Samuel Moore and Edward Aveling. Swan Sonnenschein, 1902.

———. "To Engels," June 18, 1862. In *Karl Marx and Friedrich Engels: Collected Works*, trans. Richard Dixon, 41:380. International Publishers, 1975.

Mates, Stacy G., Ivette Perfecto, and Catherine Badgley. "Parasitoid Wasp Diversity in Apple Orchards along a Pest-Management Gradient." *Agriculture, Ecosystems and Environment* 156 (2012): 82–88.

Mayr, Ernst. *The Growth of Biological Thought: Diversity, Evolution, and Inheritance*. Harvard University Press, 1982.

Mbembe, Achille. "Necropolitics." Trans. Libby Meintjes. *Public Culture* 15.1 (2003): 11–40.

———. *Out of the Dark Night: Essays on Decolonization*. Trans. Daniela Ginsburg. Columbia University Press, 2021.

McClintock, Anne. *Imperial Leather: Race, Gender, and Sexuality in the Colonial Contest*. Routledge, 2013.

McConnell, Frank. *The Science Fiction of H. G. Wells*. Oxford University Press, 1981.

Merchant, Carolyn. *The Death of Nature: Women, Ecology, and the Scientific Revolution*. Harper & Row, 1983.

Midgley, Mary. "Selfish Genes and Social Darwinism." *Philosophy* 58.225 (1983): 365–77.

Mighall, Robert. *A Geography of Victorian Gothic Fiction: Mapping History's Nightmares*. Oxford University Press, 2003.

Milburn, Colin Nazhone. "Monsters in Eden: Darwin and Derrida." *MLN* 118.3 (2003): 603–21.

Miller, David A. *Jane Austen, or The Secret of Style*. Princeton University Press, 2003.

———. *The Novel and the Police*. University of California Press, 1988.

Miller, Elizabeth Carolyn. *Extraction Ecologies and the Literature of the Long Exhaustion*. Princeton University Press, 2021.

Miller, J. Hillis, Barbara Hardy, and Richard Poirier. "*Middlemarch*, Chapter 85: Three Commentaries." *Nineteenth-Century Fiction* 35.3 (1980): 432–53.

Miller, John MacNeill. "Composing Decomposition: *In Memoriam* and the Ecocritical Undertaking." *Nineteenth-Century Contexts* 39.5 (2017): 383–98.

———. "The Ecological Plot: A Brief History of Multispecies Storytelling, from Malthus to *Middlemarch*." *Victorian Literature and Culture* 48.1 (2020): 155–85.

Mivart, George Jackson. "Darwin's *Descent of Man*." *Quarterly Review* 131 (1871): 47–90.

Moore, Burness, and Bernard Fine, eds. *A Glossary of Psychoanalytic Terms and Concepts*. American Psychiatric Association, 1968.

Moretti, Franco. *The Bourgeois: Between History and Literature*. Verso, 2013.

———. *Graphs, Maps, and Trees: Abstract Models for a Literary History*. Verso, 2005.

Morris, Francis Orpen. *All the Articles of the Darwin Faith*. W. Poole, 1882.

Morton, Timothy. *The Ecological Thought*. Harvard University Press, 2010.

———. *Ecology without Nature: Rethinking Environmental Aesthetics*. Harvard University Press, 2009.

———. *Hyperobjects: Philosophy and Ecology after the End of the World*. University of Minnesota Press, 2013.

Mulvey, Laura. "Some Thoughts on Theories of Fetishism in the Context of Contemporary Culture." *October* 65 (Summer 1993): 3–20.
Nabokov, Vladimir. *Lectures on Literature*. Ed. Fredson Bowers. Harcourt Brace Jovanovich, 1980.
Neyrat, Frédéric. "The Western Relation: The Politics of Humanism." In *The Politics of Culture*, ed. Richard Calichman and John Namjun Kim, 189–203. Routledge, 2010.
Nonini, Donald M. "Freud, Anteriority, and Imperialism." *Dialectical Anthropology* 17.1 (1992): 25–33.
Norgaard, Kari Marie. *Living in Denial: Climate Change, Emotions, and Everyday Life*. MIT Press, 2011.
Oldfield, Derek. "The Language of the Novel: The Character of Dorothea." In *Middlemarch: Critical Approaches to the Novel*, ed. Barbara Hardy, 63–85. Oxford University Press, 1967.
Oreskes, Naomi, and Erik M. Conway. *Merchants of Doubt: How a Handful of Scientists Obscured the Truth on Issues from Tobacco Smoke to Global Warming*. Bloomsbury, 2011.
Outka, Elizabeth. "The Transitory Space of *Night and Day*." In *A Companion to Virginia Woolf*, ed. Jessica Berman, 55–66. John Wiley, 2017.
Pascal, Roy. *The Dual Voice: Free Indirect Speech and Its Functioning in the Nineteenth-Century European Novel*. Manchester University Press, 1977.
Pauley, Philip J. "Samuel Butler and His Darwinian Critics." *Victorian Studies* 25.2 (Winter 1982): 161–80.
Peci, Greta. "Letter." Is This How You Feel? The Scientists, 2020. https://www.isthishowyoufeel.com/this-is-how-scientists-feel.html.
Pecora, Vincent. *Self and Form in Modern Narrative*. Johns Hopkins University Press, 1989.
Penner, Louise. "'Unmapped Country': Uncovering Hidden Wounds in *Daniel Deronda*." *Victorian Literature and Culture* 30.1 (2002): 77–97.
Peters, John G. "'To Strive, to Seek, to Find, and Not to Yield': Ulysses as Siren in Tennyson's Poem." *Victorian Review* 20.2 (1994): 134–41.
Peterson, Christopher. *Monkey Trouble: The Scandal of Posthumanism*. Fordham University Press, 2017.
Petrinovich, Lewis. *Darwinian Dominion: Animal Welfare and Human Interests*. MIT Press, 1998.
Phillips, Adam. *Darwin's Worms: On Life Stories and Death Stories*. Basic Books, 2009.
Plotz, John. "Speculative Naturalism and the Problem of Scale." *Modern Language Quarterly* 76.1 (2015): 31–56.
Plumwood, Val. *Feminism and the Mastery of Nature*. 1993. Reprint, Routledge, 2002.
Poovey, Mary. *Making a Social Body: British Cultural Formation, 1830–1864*. University of Chicago Press, 1995.
Poster, Mark. *Cultural History and Postmodernity: Disciplinary Readings and Challenges*. Columbia University Press, 1997.
Povinelli, Elizabeth. *Between Gaia and Ground: Four Axioms of Existence and the Ancestral Catastrophe of Late Liberalism*. Duke University Press, 2021.

"R. Kipling: Comparative Psychologist." *Atlantic Monthly* (June 1898): 858–59.
"Review of *Omphalos: An Attempt to Untie the Geological Knot*." *Natural History Review* 5 (1858): 55–60.
Ritvo, Lucille B. "The Impact of Darwin on Freud." *Psychoanalytic Quarterly* 43.2 (1974): 177–92.
Robbins, Bruce. *The Beneficiary*. Duke University Press, 2017.
Romanes, George John. *Mental Evolution in Man: Origin of Human Faculty*. Kegan Paul, 1888.
Romanow, Jacob. "Metafiction as Reality Effect: Trollope's Quixotism and Novel Theory." *ELH* 89.4 (2022): 1077–1105.
Ruskin, John. *The Storm Cloud of the Nineteenth Century: Two Lectures Delivered at the London Institution, February 4th and 11th, 1884*. George Allen, 1884.
Rylance, Rick. *Victorian Psychology and British Culture: 1850–1880*. Oxford University Press, 2000.
Said, Edward. *Culture and Imperialism*. Knopf, 1994.
Samalin, Zachary. "Introduction: A Map the Size of the Empire." *Criticism* 61.4 (2019): 423–42.
Saposnik, Irving S. "The Anatomy of *Dr. Jekyll and Mr. Hyde*." *Studies in English Literature* 11.4 (1971): 715–31.
Scott, Bonnie Kime. *In the Hollow of the Wave: Virginia Woolf and Modernist Uses of Nature*. University of Virginia Press, 2012.
Scott, Heidi C. M. "Industrial Souls: Climate Change, Immorality, and Victorian Anticipations of the Good Anthropocene." *Victorian Studies* 60.4 (2018): 588–610.
Seabrook, Jeremy. *The Race for Riches: The Human Cost of Wealth*. Marshall Pickering, 1988.
Sedgwick, Eve Kosofsky. *The Weather in Proust*. Duke University Press, 2011.
Shakespeare, William. *King Lear*. Signet, 1963.
Shanahan, Timothy. *The Evolution of Darwinism*. Cambridge University Press, 2004.
Shaw, George Bernard. *Back to Methuselah: A Metabiological Pentateuch*. 1922. Reprint, First World Publishing, 2007.
Sideris, Lisa. "One Step Up, Two Steps Back: Aesthetics, Ethics, and Savagery in Darwin's Theory of Evolution." *Soundings* 84.3/4 (Fall/Winter 2001): 365–88.
Sloterdijk, Peter. *Critique of Cynical Reason*. Trans. Michael Eldred. University of Minnesota Press, 1987.
Smith, Chris. "Darwin's Unsolved Problem: The Place of Consciousness in an Evolutionary World." *Journal of the History of Neuroscience* 19 (2010): 105–20.
Smith, Helen Zenna. *Not So Quiet: Stepdaughters of War*. 1930. Reprint, Feminist Press of the City University of New York, 1989.
Snell, Marilyn Berlin. "Margaret Atwood." *Mother Jones*, July/August 1997. https://www.motherjones.com/media/1997/07/margaret-atwood/.
Soper, Kate. "Greening Prometheus." In *The Greening of Marxism*, ed. Ted Benton, 81–99. Guilford, 1996.
Spencer, Herbert. *First Principles of a New System of Philosophy*. D. Appleton, 1898.
——. *The Principles of Ethics*. Vol. 1. D. Appleton, 1892.

———. *The Principles of Psychology.* Vol. 1. D. Appleton, 1873.
———. *The Principles of Psychology.* Vol. 2. Williams and Norgate, 1890.
———. *Social Statics: Or, The Conditions Essential to Human Happiness Specified, and the First of Them Developed.* J. Chapman, 1851.
Steiner, John. *Psychic Retreats: Pathological Organizations in Psychotic, Neurotic and Borderline Patients.* Routledge, 2003.
Stepan, Nancy. *Picturing Tropical Nature.* Cornell University Press, 2001.
Stevens, Wallace. *The Palm at the End of the Mind.* Ed. Holly Stevens. Vintage, 1990.
Stevenson, Robert Louis. *The Strange Case of Dr. Jekyll and Mr. Hyde.* Ed. Martin A. Danahay. 1886. Reprint, Broadview, 2005.
Stević, Aleksandar. "Convenient Cosmopolitanism: *Daniel Deronda*, Nationalism, and the Critics." *Victorian Literature and Culture* 45.3 (2017): 593–614.
Stewart-Williams, Steve. *Darwin, God, and the Meaning of Life: How Evolutionary Theory Undermines Everything You Thought You Knew.* Cambridge University Press, 2010.
Stone, Carole. "George Eliot's *Daniel Deronda:* The Case-History of Gwendolen H." *Nineteenth Century Studies* 7 (1993): 57–67.
Straley, Jessica. *Evolution and Imagination in Victorian Children's Literature.* Cambridge University Press, 2016.
Sutherland, Alexander. *Victoria and Its Metropolis: Past and Present.* McCarron, Bird, 1888.
Suvin, Darko. "Wells as the Turning Point of the SF Tradition." *Minnesota Review* 4.1 (1975): 106–15.
Swartz, Sally. *Psychoanalysis and Colonialism: A Contemporary Introduction.* Routledge, 2023.
Szabados, Béla. "Freud, Self-Knowledge and Psychoanalysis." *Canadian Journal of Philosophy* 12.4 (1982): 691–707.
Taylor, Charles. *The Sources of the Self: The Making of the Modern Identity.* Cambridge University Press, 1992.
Taylor, Jesse Oak. "Darwin after Nature." In *After Darwin: Literature, Theory, and Criticism in the Twenty-First Century*, ed. Devin Griffiths and Deanna Kreisel, 19–32. Cambridge University Press, 2022.
———. *The Sky of Our Manufacture: The London Fog in British Fiction from Dickens to Woolf.* University of Virginia Press, 2016.
———. "Wilderness after Nature: Conrad, Empire and the Anthropocene." In *Conrad and Nature*, ed. Lissa Schneider-Rebozo, Jeffrey Mathes McCarthy, and John G. Peters, 19–42. Routledge, 2018.
Tennyson, Alfred Lord. *The Major Works.* Ed. Adam Roberts. Oxford University Press, 2000.
———. "The Making of Man." In *The Death of Oenone, Akbar's Dream, and Other Poems*, 85–86. Macmillan, 1892.
Terry, Reginald. *Anthony Trollope: The Artist in Hiding.* Rowman and Littlefield, 1977.
Thompson, E. P. "Socialist Humanism: An Epistle to the Philistines." *New Reasoner* 1 (Summer 1957): 122.
Thompson, Janna L. "The New Social Darwinism: The Politics of Sociobiology." *Politics* 17.1 (1982): 121–28.

Tolstoy, Leo. *The Death of Ivan Ilyich and Other Stories*. Trans. Richard Peavear and Larissa Volokhonsky. Knopf, 2009.
Tondre, Michael. "George Eliot's 'Fine Excess': *Middlemarch*, Energy, and the Afterlife of Feeling." *Nineteenth-Century Literature* 67.2 (2012): 204–33.
"Treasure Houses." *The Academy* 88 (March 27, 1915): 199.
Trollope, Anthony. *The Last Chronicle of Barset*. 1867. Reprint, Knopf, 1909.
———. *Thackeray*. Macmillan, 1879.
Tromanhauser, Vicki. "Mrs. Dalloway's 'Animals and the Humanist Laboratory.'" *Twentieth-Century Literature* 58.2 (2012): 187–212.
Trunnell, Eugene E., and William E. Holt. "The Concept of Denial or Disavowal." *Journal of the American Psychoanalytic Association* 22.4 (1974): 769–84.
Tuan, Yi-Fu. *Escapism*. Johns Hopkins University Press, 1998.
Tucker, Herbert. "The Fix of Form: An Open Letter." *Victorian Literature and Culture* 27.2 (1999): 531–35.
Tyndall, John. *Heat Considered as a Mode of Motion*. Longman, Green, 1863.
"UN Chief Warns against 'Sleepwalking to Climate Catastrophe.'" *UN News*, March 21, 2022. https://news.un.org/en/story/2022/03/1114322.
Vandertop, Caitlin. "'The Earth Seemed Unearthly': Capital, World-Ecology, and Enchanted Nature in Conrad's *Heart of Darkness*." *Modern Fiction Studies* 64.4 (2018): 680–700.
Wald, Priscilla. "Science and Technology." In *A Companion to Critical and Cultural Theory*, ed. Imre Szeman, Sarah Blacker, and Justin Sully, 403–18. Routledge, 2017.
Walker, Richard J. *Labyrinths of Deceit: Culture, Modernity and Identity in the Nineteenth Century*. Liverpool University Press, 2007.
Wallace, Alfred Russel. *Darwinism: An Exposition of the Theory of Natural Selection, with Some of Its Applications*. Macmillan, 1890.
———. *The Malay Archipelago*. Macmillan, 1872.
———. "The Origin of the Human Races and the Antiquity of Man Deduced from the Theory of Natural Selection." *Journal of the Anthropological Society of London* 2 (1864): clviii–clxxxvii.
———. *The Wonderful Century: The Age of New Ideas in Science and Invention*. Allen, 1908.
Ward, Arthur D. "'Ulysses' and 'Tithonus': Tunnel-Vision and Idle Tears." *Victorian Poetry* 12.4 (1974): 311–19.
Weismann, August. "The All-Sufficiency of Natural Selection: A Reply to Herbert Spencer." *Contemporary Review* 64 (September 1893): 309–38.
Wells, H. G. *The Island of Dr. Moreau*. Ed. Steve Maclean and Patrick Parrinder. 1896. Reprint, Penguin, 2005.
———. *The Time Machine*. Ed. Nicholas Ruddick. 1895. Reprint, Broadview, 2001.
Whitebook, Joel. *Freud: An Intellectual Biography*. Cambridge University Press, 2017.
Wilberforce, Samuel. "Darwin's *Origin of Species*." *Quarterly Review* 108 (July and October, 1860): 225–64.
Williams, Raymond. *Keywords: A Vocabulary of Culture and Society*. Oxford University Press, 1983.

Wilson, Edward O. "Foreword." In *Common Wealth: Economics for a Crowded Planet*, by Jeffrey Sachs, xi–xiii Penguin, 2008.
———. *The Future of Life*. Vintage, 2003.
———. *On Human Nature*. 1978. Reprint, Harvard University Press, 2012.
Wood, James. *How Fiction Works*. Farrar, Straus and Giroux, 2008.
Woolf, Virginia. *Between the Acts*. Harcourt Brace, 1941.
———. *The Collected Essays of Virginia Woolf*. Vol. 2. Hogarth Press, 1966.
———. *The Common Reader First Series*. Ed. Andrew McNeillie. Harcourt Brace, 1984.
———. *The Diary of Virginia Woolf*. Vol. 1, *1915–1919*. Ed. Anne Olivier Bell. Houghton Mifflin, 1977.
———. *The Diary of Virginia Woolf*. Vol. 2, *1920–1924*. Ed. Anne Olivier Bell. Houghton Mifflin, 1978.
———. *Jacob's Room*. 1922. Reprint, Dover, 1998.
———. *The Letters of Virginia Woolf*. Vol. 2, *1912–1922*. Ed. Nigel Nicolson and Joanne Trautmann. Harcourt Brace Jovanovich, 1976.
———. *Mrs. Dalloway*. 1925. Reprint, Harvest, 1981.
———. "On Being Ill." *New Criterion* (January 1926): 32–45.
———. *To the Lighthouse*. 1927. Reprint, Harvest, 1981.
———. *The Voyage Out*. Duckworth, 1920.
Worster, Donald. *Nature's Economy: A History of Ecological Ideas*. Cambridge University Press, 1994.
Wynter, Sylvia. "Unsettling the Coloniality of Being/Power/Truth/Freedom: Towards the Human, after Man, Its Overrepresentation—an Argument." *CR: The New Centennial Review* 3.3 (2003): 257–337.
Yeats, William Butler. *The Collected Poems of W. B. Yeats*. Ed. Richard Finneran. Scribner, 1996.
Zalloua, Zahi. *Being Posthuman: Ontologies of the Future*. Bloomsbury, 2020.
Zhang, Dora. "Naming the Indescribable: Woolf, Russell, James, and the Limits of Description." *New Literary History* 45.1 (2014): 51–70.
Zilcosky, John. "Savage Science: Primitives, War Neurotics, and Freud's Uncanny Method." *American Imago* 70.3 (2013): 461–86.
Žižek, Slavoj. *For They Know Not What They Do: Enjoyment as a Political Factor*. Verso, 2008.
———. "Introduction: The Spectre of Ideology." In *Mapping Ideology*, ed. Slavoj Žižek, 1–33. Verso, 1994.

INDEX

Achebe, Chinua, 117
Adorno, Theodor, 75, 219
Albrecht, Thomas, 178
Alice in Wonderland (Carroll), 19, 38, 110
Allen, Amy, 89–90, 95
All Quiet on the Western Front (Remarque), 211
Anderson, Amanda, 177–78
Anderson, Kay, 63, 103–4, 190
Anthropocene, 10–11, 66, 109, 207, 214
anthropogenic climate change: commodity fetishism and, 44–45; denial and, 4, 18, 40–42, 45, 212–13; disavowal and, 2, 39, 48, 109, 131–32, 215, 225, 227–28; as hyperobject, 22; misinformation regarding, 2–3; oil and gas industry and, 18
Appiah, Kwame Anthony, 178
Arata, Stephen, 114
Arendt, Hannah, 49
Atwood, Margaret. *See also specific works*: Darwin and, 223; *Wuthering Heights* and, 218
Austen, Jane, 35, 140, 142

Bagehot, Walter, 80–81
Bakhtin, M. M., 136
Bateson, Gregory, 31
Becker, Ernest, 4–5, 30
Beer, Gillian, 27, 76, 135, 187
bees, 233–35, 257nn55–56
Benjamin, Walter, 35, 48–49
Berlant, Lauren, 39–40
Between the Acts (Woolf), 187–91
Bilgrami, Akeel, 33–34
Bleak House (Dickens), 70, 181
Booth, Wayne, 136
Bouissac, Paul, 238n23
Bowen, Peter, 142
Bowler, Peter, 72–73, 123, 128
Brontë, Charlotte, 28–29, 39–40, 42, 227
The Brothers Karamazov (Dostoevsky), 25, 223
Browne, Janet, 36–37, 52
Buffon, Comte de (George LeClerc), 59
Butler, Samuel, 72, 74

capitalism: climate crisis and, 18; commodity fetishism and, 42–43; denaturalization and, 33; denial and, 5–6, 50, 230; evolutionary biology and, 18–19, 48, 64; extinction of species and, 68; Great War and, 187
Carroll, David, 134, 152

Carson, Rachel, 11–12, 56, 235
Castoriadis, Cornelius, 94
Chaitin, Gilbert, 138
Chakrabarty, Dipesh, 10–11, 17, 161, 218–19
Challenger, Melanie, 99
Chambers, Robert, 58
A Christmas Carol (Dickens), 110
The City of Dreadful Night (Thomson), 181
Civilization and Its Discontents (Freud), 89–91, 93
Clark, Christopher, 213
Clark, Timothy, 35, 47, 198, 212
Clifford, William Kingdon, 29, 214
climate crisis. *See* anthropogenic climate change
Cohen, Stanley, 2, 119, 144, 149, 171
Cohn, Dorrit, 136
colonialism. *See* imperialism
commodity fetishism, 42–45
Conrad, Joseph. *See Heart of Darkness* (Conrad)
Cope, Edward Drinker, 72
Cyrus the Great, 160–61

Daniel Deronda (Eliot): denial and, 169–71, 174–76; Deronda's racial identity in, 174–76, 178–79; disavowal and, 179–80; ethnic nationalism in, 177–79; free indirect discourse in, 166, 171–72, 175; imperialism and, 164–65; multiplot narrative structure of, 133; vision of history in, 163–64, 177; Zionism and, 174, 177–78
Darwin, Charles. *See also* evolutionary biology; natural selection; *specific works*: anti-teleology of, 6; Atwood and, 223; denial personally experienced and expressed by, 1, 52; ecotheory and, 13; Eliot and, 134–36, 158, 160; environmentalism and, 67; Fabre and, 232–34; Kipling and, 128–29; Moor Park hydrotherapy retreat and, 36, 51–52, 54, 96; novel reading by, 36–37; physical health of, 52; Tierra del Fuego journey of, 84, 85; on ubiquity of extinction, 52–54, 135; Woolf and, 16, 187, 190, 254n20; Wordsworth and, 218
David Copperfield (Dickens), 220–21
Dawkins, Richard, 6
The Death of Ivan Ilych (Tolstoy), 138–39
The Death of Nature (Merchant), 30
Deer, Jemma, 90
Deese, R. S., 214
denial. *See also* hard denial; soft denial: advertising and, 43–44; anthropogenic climate change and, 4, 18, 40–42, 45, 212–13; the body and, 5, 75, 86; capitalism and, 5–6, 50, 230; conscious self-awareness and, 99; denied dependency and, 43, 203–4, 206; deviant criteria of assent and, 19–20; evolutionary biology and, 14–15, 18–22, 25–26, 35, 41, 55, 57, 60, 62–63, 75, 84, 120, 215, 223, 226, 233; free indirect discourse and, 132, 134, 135, 138–43, 145, 147, 149, 151; Freudian psychoanalysis and, 87–89, 93, 96; Great War and, 17, 187, 209–12; imperialism and, 5, 30–31, 120, 125, 185; implicatory denial and, 2; interpretive denial and, 149, 171; ironic self-consciousness and, 46; knowing and, 45; liberalism and, 32, 40–41; Marxist critique and, 43; mortality and, 4; opposite of, 228; repression and, 87; science denialism and, 22, 30–31; self-denial and, 75; *To The Lighthouse* and, 16–17, 184, 186–87, 195–98, 201, 203, 205–6, 214–16; Victorians and, 14, 16, 22, 41, 46, 184
The Descent of Man (Darwin): biological egalitarianism in, 32; comparison between nonhuman animals and "savages" in, 84; Homo duplex and, 59–60; imperialism and, 31–32, 158; kinship questions in, 83–84; Mivart's review of,

76–77; self-consciousness and, 82–83; sympathy beyond the species barrier and, 158; white supremacy and, 31–32, 83–84, 158

Desmond, Adrian, 61

Dickens, Charles: *Bleak House* and, 70, 181; *A Christmas Carol* and, 110; *David Copperfield* and, 220–21; *Great Expectations* and, 98; *Martin Chuzzlewit* and, 207; *Our Mutual Friend* and, 70, 207–8

Dickinson, Janis, 4

Didion, Joan, 50

Dillard, Annie. *See Pilgrim at Tinker Creek* (Dillard)

disavowal: anthropogenic climate change and, 2, 39, 48, 109, 131–32, 215, 225, 227–28; artfulness of, 34–35; civilization and, 50, 91–92; definition of, 2; as "denial of denial," 46; evolutionary biology and, 29, 41, 180; extinction and, 39, 54, 222; fetishism and, 44, 87; Freudian psychoanalysis and, 87–89, 92–93; immortality ideology and, 4; liberalism and, 32, 39, 57; natural selection and, 27; realism and, 15, 35, 39; self-deception and, 34–35; willing suspension of disbelief and, 34

Dodds, Joseph, 119

Dostoevsky, Fyodor, 25, 223

Dr. Jekyll and Mr. Hyde (Stevenson): contradictory attitudes toward scientific experimentation in, 112; racialized identities of Jekyll and Hyde in, 114–15; self-consciousness and, 113–15; shifts from third-person to first-person narration in, 111, 112–13, 114

Duncan, Ian: on *Daniel Deronda*, 165, 168–69, 180; on Darwin and sympathy beyond the species barrier, 158; on free indirect discourse, 136; on "Homo duplex," 59–60; on the nineteenth-century novel as zone of experimentation, 224–25

Eliot, George. *See also specific works*: Darwin and, 134–36, 158, 160; denial and, 24–26, 40, 133, 135–36, 157, 159–60, 185; epistemology and ethics in the works of, 153; historicism and, 162–63; human exceptionalism and, 160; liberalism and, 25, 39; natural selection and, 27, 135; realism and, 133, 160

Ellegård, Alvar, 71–72

Emma (Austen), 140, 142

empire. *See* imperialism

enlightened false consciousness, 44

The Enlightenment: Darwin and reevaluation of, 13; denial and, 5; human exceptionalism and, 15, 227; *Oryx and Crake* as satire of, 227; rationality and, 90; secularism and, 95; self-consciousness and, 94

eugenics, 86

Ever Since Darwin (Gould), 6

Evidence as to Man's Place in Nature (Huxley), 61, 197–98

Evolution and Ethics (Huxley), 65

evolutionary biology. *See also* natural selection: bioengineering and, 219; capitalism and, 18–19, 48, 64; denial of, 14–15, 18–22, 25–26, 35, 41, 55, 57, 60, 62–63, 75, 84, 120, 215, 223, 226, 233; disavowal of, 29, 41, 180; human exceptionalism and, 14, 15, 29, 58, 61, 68, 71, 86, 103–4, 157, 214; humans decentered in, 12, 33–34, 68, 73–74, 98–99, 135; as hyperobject, 22; imperialism and, 88, 120; liberalism and, 26, 32; Lyellian geology and, 36; natural history and, 161, 188; refusal to accept metaphysical and moral consequences of, 21–22; teleology and, 30, 33; Victorians and, 14, 20, 22; white supremacy and, 64

evolutionary psychology, 76–77

extinction: capitalism and, 68; Darwin on the ubiquity of, 52–54, 135; disavowal

extinction (*cont.*)
  and, 39, 54, 222; human anxiety regarding, 24–25, 100; *In Memoriam* and, 100; *Middlemarch* and, 157; soft denial and, 15, 27–28; *The Time Machine* and, 108, 110, 120; Victorians and, 40, 182, 184

Fabian, Johannes, 84–85, 95
Fabre, Jean-Henri, 232–34, 256n52
Facing Gaia (Latour), 2, 212
Fanon, Frantz, 32
Ferguson, Frances, 136–37, 142
Feser, Ali, 221
Finch, Casey, 142
First World War. *See* Great War
Flammarion, Camille, 181
Flatland (Abbot), 110
Flaubert, Gustave. *See Madame Bovary* (Flaubert)
Fowles, John, 5
Freedgood, Elaine, 138, 141–42
free indirect discourse: *Daniel Deronda* and, 166, 171–72, 175; denial and, 132, 134, 135, 138–43, 145, 147, 149, 151; as "dual voice," 132, 136–37, 143–44; fetishistic nature of language and, 148; gap between character and world in, 139; *Jude the Obscure* and, 142; *Middlemarch* and, 145–49, 151, 153–54, 157–58; *Mrs. Dalloway* and, 191–92; *Oryx and Crake* and, 220; syntatic ambiguity and, 220; third-person component of, 16, 137, 141–42; *To the Lighthouse* and, 195, 199, 202; uncertainty of perspective in, 147
Freud, Anna, 87
Freud, Sigmund. *See also specific works*: animism and, 90; civilization and, 89–94; Darwin and, 92; denial and, 87–89, 93, 96; disavowal and, 87–89, 92–93; Enlightenment values and, 90, 95–96; extractive capitalism and, 90; fetishism and, 44; human exceptionalism and, 91; imperialism and, 88–89, 94; Oedipus complex and, 95; on religious belief and human weakness, 89
Full House (Gould), 73
The Future of an Illusion (Freud), 89

Gallagher, Catherine, 154
Genette, Gerard, 136
Ghosh, Amitav, 35–37, 116, 131–33
Glasson, Ben, 45
Goblin Market (Rossetti), 110
Gosse, Phillip, 19
Gould, Stephen Jay, 6–7, 32, 56, 58, 73, 102
The Grammar of Science (Pearson), 34
Gray, Asa, 73, 234
Gray, John, 99
The Great Derangement (Ghosh), 35, 116, 131–32
Great Expectations (Dickens), 98
Great War: denial and, 17, 187, 209–12; environmental history and, 187–88, 214; noncombatants and, 210–11; *To the Lighthouse* and, 186, 208–9, 215; Woolf and, 209–12, 214
Griffiths, Devin, 13
Grosz, Elizabeth, 4, 13–14
Grove, Richard, 67
Gulliver's Travels (Swift), 226–27
Guterres, António, 48

Haeckel, Ernst, 71–72, 77–78, 128–29
Hale, Piers, 63
Hansson, Sven Ove, 19
Haraway, Donna, 6, 14, 234
hard denial: anthropogenic climate change and, 18; evolutionary biology and, 18–19, 22, 25, 60; soft denial normalized through, 13, 60
Hardy, Thomas. *See Jude the Obscure* (Hardy)
Hartman, Saidiya, 158–59
Heart of Darkness (Conrad): as climate change novel, 117; denial and disavowal in, 106, 115, 118, 119, 185; en-

vironmental imaginary and, 117; final scene in, 119; imperialism and, 117, 119, 125, 185; Kurtz's Intended in, 106, 119, 185; landscape description in, 116; racial undercurrents in representation of self-consciousness in, 115
heat death of the sun, 108, 181, 184
Hegel, G.W.F., 245–46n102
Herbert, Christopher, 33–34
Herodotus, 161
hippocampus minor controversy, 62
Holt, William, 87, 93
Homo duplex, 59–60
Horkheimer, Max, 75, 219
Horton, Zach, 197
human exceptionalism: The Enlightenment and, 15, 227; evolutionary biology and, 14, 15, 29, 58, 61, 68, 71, 86, 103–4, 157, 214; grounds of differentiation between humans and non-human animals and, 61–63, 82; historicism and, 162–63; human origin of ethics and, 63; Lamarckism and, 72, 74, 129; liberalism and, 40; self-conscious reflection and, 16, 69, 76–79, 82–83, 86, 99, 114; Tennyson and, 103–4; white supremacy and, 10, 31, 80, 83–84, 114, 126, 160
Human Forms (Duncan), 38
Hustak, Carla, 234–35
Hutton, James, 26
Huxley, Julian, 85–86
Huxley, Thomas. *See also specific works*: civilization viewed as stronghold against violence in nature by, 63, 65–66, 115; European superiority asserted by, 65; grounds of differentiation between humans and non-human animals and, 61–63; hippocampus minor controversy and, 62; human origin of ethics for, 63; *The Origin of Species* reviewed by, 60; scientific humanism and, 61
hyperobjects, 22–23, 36

ichneumon wasps, 234–35
immortality ideology (Becker), 4
imperialism: Anthropocene and, 11; civilizational rhetoric and, 66, 80, 125; denial and, 5, 30–31, 120, 125, 185; environmental imaginary and, 117; evolutionary biology and, 88, 120; Freudian psychoanalysis and, 88–89, 94; Lamarckism and, 81, 125; Victorians and, 41, 117; white supremacy and, 30–32, 81, 114, 125, 158, 179
The Impressions of Theophrastus Such (Eliot), 162–63
In Memoriam (Tennyson): denial and disavowal in, 102–3, 106–7, 118, 122; elegy genre and, 101; extinction and, 100; human exceptionalism and, 103–4; indeterminacy of belief system in, 105–7; nature and violence in, 23; scientific naturalism and, 101; veil tropes in, 26–27
interpretive denial, 149, 171
The Island of Dr. Moreau (Wells): ambiguity in ending of, 121, 125; beautiful soul syndrome and, 127; denial and disavowal in, 120, 122, 124–26; escape and, 122; evolutionary biology and, 120, 123, 126; human exceptionalism and human consciousness in, 120–21, 125–26, 129–30; imperialism and, 120, 122, 125, 127–28; *The Jungle Book* and, 127–28; Lamarckism and, 123–26

Jackson, Zakiyyah Iman, 83, 85, 158–59
Jacob's Room (Woolf), 192–93, 194, 254n24
James, Henry: on denial and artistic practice, 193–94; *Partial Portraits* and, 165; *Roderick Hudson* and, 193; *Washington Square* and, 140–41
Jameson, Fredric, 39, 99–100, 106, 116, 227
Journal of Researches (Darwin), 36, 52, 55, 85, 187
Jude the Obscure (Hardy), 138–39, 142–44
The Jungle Book (Kipling), 127–29, 249n84

Kant, Immanuel, 244–45n75
Karera, Axelle, 11
Keith, Arthur, 71
Khanna, Ranjana, 80, 88
King Lear (Shakespeare), 125
Kingstone, Helen, 160, 162
Kipling, Rudyard, 127–29. See also *The Jungle Book* (Kipling)
Klein, Naomi, 2, 13, 27, 39, 237n7
Knoll, Elizabeth, 83
Kohn, David, 57
Kreilkamp, Ivan, 154–55
Kucich, John, 135, 157, 230

Lamarckism: antimaterialism and, 72–73; denial and, 125; human exceptionalism and, 72, 74, 129; imperialism and, 81, 125; *The Island of Dr. Moreau* and, 123–26; Kipling and, 128; monetary metaphors and, 82; plasticity and, 123–24; self-consciousness and, 129; Spencer and, 72, 74, 81–82; teleology and, 72; white supremacy and, 81
Latour, Bruno, 2, 6, 14, 90, 187, 212–13
"Leaves from a Notebook" (Eliot), 25
Leavis, F. R., 116, 172, 177
Letters on the Laws of Man's Nature and Development (Martineau), 28–29
Levenback, Karen, 209–10
Levine, George, 22–23, 61, 152–54, 241n42
liberalism: climate crisis and, 40–41; denial and, 32, 40–41; disavowal and, 32, 39, 57; evolutionary biology and, 26, 32; human exceptionalism and, 40; liberal humanism and, 6, 40–41; teleology and, 32–33, 40, 57, 221; white supremacy and, 83
The Lifted Veil (Eliot), 24–25
Lilith (MacDonald), 110
Lubbock, John, 88
Lyell, Charles, 26, 76, 197

MacBeth (Shakespeare), 205
Madame Bovary (Flaubert), 138–39
MaddAddam (Atwood), 230
MaddAddam trilogy (Atwood). See *MaddAddam* (Atwood); *Oryx and Crake* (Atwood); *The Year of the Flood* (Atwood)
"The Making of Man" (Tennyson), 29–30, 103
The Malay Archipelago (Wallace), 68–69
Malm, Andreas, 18
Malthus, Thomas, 53
Mannoni, Octave, 2, 34, 43, 45–46, 60, 189
Marcuse, Herbert, 93
Martin Chuzzlewit (Dickens), 207
Martineau, Harriet, 28–29, 42, 227
Marx, Karl, 19, 42–43, 162, 239n2
Maupassant, Guy de, 35
Mayr, Ernst, 59
Mbembe, Achille, 126, 230
McClintock, Anne, 88
McConnell, Frank, 120
Mental Evolution in Man (Romanes), 77
Merchant, Carolyn, 30
Middlemarch (Eliot): backgrounding of the non-human world in, 156; denial and, 138, 144–48, 150–56, 174–75; extinction and, 157; free direct discourse in, 145–49, 151, 153–54, 157–58; the gothic and, 151–53; multi-plot narrative structure of, 133; vision of history in, 160–61
Miller, D. A., 136–37, 140, 142
Miller, Elizabeth Carolyn, 164
Miller, J. Hillis, 149–51, 159
Miller, John MacNeill, 102, 158
The Mill on the Floss (Eliot), 141, 155–57, 161
Mivart, St. George Jackson, 76–77, 96
Moretti, Franco, 27, 134, 136
Morris, Francis, 240n12
Morris, William, 79
Morton, Timothy: on beautiful soul syndrome, 127; on Darwin as deconstructionist, 13; on difficulty of ecological

thought, 13, 187; on the Great War, 213–14; on hyperobjects, 22–23
Mrs. Dalloway (Woolf): Clarissa's reading of Victorian scientific naturalists in, 188–89; disavowal and, 189; free indirect discourse in, 191–92
Mulvey, Laura, 43–44
Myers, Natasha, 234–35

Nabokov, Vladimir, 35
naturalism, 97–98, 156
natural selection. *See also* evolutionary biology: denial and, 20; disavowal and, 27; Eliot and, 27, 135; extinction and, 54; Ichneumonidae wasps and, 234–35; persistence of "lower organisms" and, 20, 71; random variation and, 72–73; units of survival and, 31; violence and, 23
Neo-Lamarckism. *See* Lamarckism
Neyrat, Frédéric, 81
Nonini, Donald, 88
Norgaard, Kari Marie, 2, 8, 108
Not So Quiet: Stepdaughters of War (Price), 210–11

Oedipus Rex (Sophocles), 121, 126, 225–26
Oldfield, Derek, 145–46
Omega: The Last Days of the World (Flammarion), 181
Omphalos (Gosse), 19
On Being Ill (Woolf): the body and, 199; denial and, 183–84; entropy and, 182; knowledge and, 181
On Human Nature (Wilson), 7–8
On The Physical Basis of Life (Huxley), 156
"The Origin of Human Races" (Wallace), 162
The Origin of Species (Darwin): bees in, 234, 257nn55–56; checks to population growth and, 53; denial and, 1, 15, 52, 55; on evolution and time horizons, 22–23; on "the face of Nature," 54–55; human ecological interventions downplayed in, 67–68; Huxley's review of, 60; Ichneumonidae wasps and, 234; natural selection and, 22–23; on the ubiquity of extinction, 54, 57; vision of history in, 161; Wilberforce's review of, 20
Oryx and Crake (Atwood): agency questions in, 220, 224; bildungsroman conventions and, 218, 221; bioengineering and, 229, 232; biological reproduction and, 222–24; denial and, 222–23, 228; disavowal and, 220; The Enlightenment satirized in, 227; free indirect discourse and, 220; *Gulliver's Travels* epigraph in, 226–27; self-consciousness and, 222; *To the Lighthouse* epigraph in, 227; transhumanism and, 223
Our Mutual Friend (Dickens), 70, 207–8
Outka, Elizabeth, 210
An Outline of Psychoanalysis (Freud), 87
Owen, Richard, 62

Partial Portraits (James), 165
Pascal, Roy, 132, 139–40, 141, 147
Pauley, Philip, 73
Pearson, Karl, 33–34
Peci, Greta, 42
Pecora, Vincent, 136
Penner, Louise, 170
Peterson, Christopher, 92
Petrinovich, Lewis, 73
Phillips, Adam, 51–52, 57–58, 109
Physics and Politics (Bagehot), 80
Pilgrim at Tinker Creek (Dillard), 1, 3, 227, 232–33, 235–36
Pilgrim's Progress (Bunyan), 111
Plotz, John, 97
Plumwood, Val, 43, 156, 203–4, 206
Poovey, Mary, 47–48
Poster, Mark, 162
Povinelli, Elizabeth, 32, 57, 221
Price, Evadne, 93, 110, 210–11
The Principles of Ethics (Spencer), 78–79
The Principles of Psychology (Spencer), 78, 80
Prometheus, 7

racism. *See* white supremacy
realism: catastrophic possibilities obscured in, 35–37, 131; disavowal and, 15, 35, 39; environmental denialism and, 131; free indirect discourse and, 132, 136; Ghosh's criticisms of, 35–37, 116, 131; willing suspension of disbelief and, 34, 38
Remarque, Erich, 211
Robbins, Bruce, 44
Roderick Hudson (James), 193
Romanes, George, 77–78, 80
Romanow, Jacob, 37
Ruskin, John, 70, 79, 206
Rylance, Rick, 77, 81–82

Said, Edward, 118
Samalin, Zachary, 46–48, 88
Saposnik, Irving, 112
scalar collapse, 197–98
scientific humanism, 8, 61
scientific materialism, 7, 9, 100, 106, 227
scientific naturalism, 101, 108, 182, 187–88
Scott, Heidi, 214
Seabrook, Jeremy, 43–44
Sedgwick, Eve Kosofsky, 3–4, 63
Shaw, George Bernard, 74, 78, 128
The Sleepwalkers (Clark), 213
Sloterdijk, Peter, 44–45
Social Darwinism, 10, 63–65, 72, 80
soft denial: evolutionary biology and, 15, 25–26, 35, 60, 62–63, 223; extinction and, 15, 27–28; hard denial as means of normalizing, 13, 60; self-deception and, 28, 37, 46–47, 49; veil tropes and, 26; voice and, 27
Soper, Kate, 42–44
Souvenirs entomologiques (Fabre), 232–33, 256n52
Spencer, Herbert: on desire and self-consciousness, 78–79; empire and, 81; human exceptionalism and, 79; laissez-faire economic individualism and, 66, 79; Lamarckism and, 72, 74, 81–82; on mental revolutions and "laceration," 14, 74–75; on self-consciousness and human nature, 78–79
Steiner, John, 34
Stepan, Nancy, 67–68
Stevens, Wallace, 101, 194–95
Stevenson, Robert Louis. *See Dr. Jekyll and Mr. Hyde* (Stevenson)
Stević, Aleksandar, 178–79
Stewart-Williams, Steven, 73
Stone, Carole, 170
The Storm Cloud of the Nineteenth Century (Ruskin), 70
Suvin, Darko, 127–28
Swartz, Sally, 88
Swift, Jonathan, 226–27

Taylor, Charles, 8, 75, 228–29, 231
Taylor, Jesse Oak, 116, 207, 241n58
Tennyson, Alfred Lord: denial and, 40, 102–3, 106–7, 118, 122; human exceptionalism and, 103–4; "The Making of Man" and, 29–30, 103; *In Memoriam* and, 26, 100–102, 104–8, 118, 122; "Ulysses" and, 104–5, 122, 197; veil tropes and, 26–27, 29, 222
Terry, Reginald, 36
Thompson, E. P., 162
Thomson, James, 181
Through the Looking Glass (Carroll), 110
The Time Machine (Wells): apocalypse in, 110–11, 181; disavowal and, 108–10, 118; human extinction and, 108, 110, 120
Tolstoy, Leo. *See The Death of Ivan Ilych* (Tolstoy)
Totem and Taboo (Freud), 88, 94–95
To the Lighthouse (Woolf): Anthropocene and, 207; death of Andrew Ramsay in, 209; death of Mrs. Ramsay in, 204–5; denial and, 16–17, 184, 186–87, 195–98, 201, 203, 205–6, 214–16; dinner scene in, 201–2; entropy and, 205, 207; free indirect discourse and, 195, 199, 202; Great War and, 186, 208–9, 215; *Oryx and Crake*'s epigraph from, 227; scalar

collapse and, 186, 197–98; the sea in, 194–95; servants in, 205
transhumanism, 86, 223
Trollope, Anthony, 35–38
Tromanhauser, Vicki, 189
Trunnell, Eugene, 87
Tuan, Yi-Fu, 5, 30
Tucker, Herbert, 105–6
Tylor, Edward, 88
Tyndall, John, 76, 188

"Ulysses" (Tennyson), 104–5, 122, 197, 247n19
unreliable narration, 98
Unto This Last (Ruskin), 70

Vandertop, Caitlin, 116
veil tropes: Brontë (Charlotte) and, 28–29; commodity fetishism and, 42–43; Dillard and, 233; Eliot and, 26–27; extent of concealment and, 26; soft denial and, 26; Tennyson and, 26–27, 29, 222
Vestiges of the Natural History of Creation (Chambers), 58
The Voyage of the Beagle (Darwin), 36, 52, 55, 85, 187
The Voyage Out (Woolf), 187, 196–97

Wald, Priscilla, 85–86
Wallace, Alfred Russel, 68–71, 162, 244n53
The War of the Worlds (Wells), 122
Washington Square (James), 140–41
The Weather in Proust (Sedgwick), 3
Weismann, August, 129
Wells, H.G., 129, 130, 185. *See also specific works*

"What Is Freedom?" (Arendt), 49
Whitebook, Joel, 94
white supremacy: civilization and, 64; climate crisis and, 11; eugenics and, 86; evolutionary biology and, 64; human exceptionalism and, 10, 31, 80, 83–84, 114, 126, 160; imperialism and, 30–32, 81, 114, 125, 158, 179; Lamarckism and, 81; liberalism and, 83; self-consciousness and, 80–81
Wilberforce, Samuel, 20, 22–23
Williams, Raymond, 111
Wilson, E. O., 7–11
The Wonderful Century (Wallace), 69
Wood, James, 149, 199, 251n54
Woolf, Virginia. *See also specific works*: Darwin and, 16, 187, 190, 254n20; denial and, 16, 184–86, 191, 194; disavowal and, 186; Great War and, 209–12, 214; on her feminism, 211; on realism, 35; scientific naturalism and, 182, 187–88
Wordsworth, William, 210, 218
World War I. *See* Great War
Worster, Donald, 64, 117
Wuthering Heights (Emily Brontë), 38, 110, 218
Wynter, Sylvia, 10–11, 30, 32, 126–27

The Year of the Flood (Atwood), 217, 223, 230–31
Yeats, William Butler, 111, 182

Zalloua, Zahi, 86
Zhang, Dora, 114
Zilcosky, John, 94–95
Žižek, Slavoj, 44–45, 86

The authorized representative in the EU for product safety and compliance is:
Mare Nostrum Group
B.V Doelen 72
4831 GR Breda
The Netherlands

www.ingramcontent.com/pod-product-compliance
Lightning Source LLC
Chambersburg PA
CBHW031800220426
43662CB00007B/477